Visual Thinking in Mathematics

Visual Thinking in Mathematics

An epistemological study

M. Giaquinto

OXFORD
UNIVERSITY PRESS

Great Clarendon Street, Oxford OX2 6DP

Oxford University Press is a department of the University of Oxford.
It furthers the University's objective of excellence in research, scholarship,
and education by publishing worldwide in

Oxford New York

Auckland Cape Town Dar es Salaam Hong Kong Karachi
Kuala Lumpur Madrid Melbourne Mexico City Nairobi
New Delhi Shanghai Taipei Toronto

With offices in

Argentina Austria Brazil Chile Czech Republic France Greece
Guatemala Hungary Italy Japan Poland Portugal Singapore
South Korea Switzerland Thailand Turkey Ukraine Vietnam

Oxford is a registered trade mark of Oxford University Press
in the UK and in certain other countries

Published in the United States
by Oxford University Press Inc., New York

© Marcus Giaquinto 2007

The moral rights of the authors have been asserted
Database right Oxford University Press (maker)

First published 2007

All rights reserved. No part of this publication may be reproduced,
stored in a retrieval system, or transmitted, in any form or by any means,
without the prior permission in writing of Oxford University Press,
or as expressly permitted by law, or under terms agreed with the appropriate
reprographics rights organization. Enquiries concerning reproduction
outside the scope of the above should be sent to the Rights Department,
Oxford University Press, at the address above

You must not circulate this book in any other binding or cover
and you must impose the same condition on any acquirer

British Library Cataloguing in Publication Data
Data available

Library of Congress Cataloging in Publication Data
Data available

Typeset by Laserwords Private Limited, Chennai, India
Printed in Great Britain
on acid-free paper by
Biddles Ltd., King's Lynn, Norfolk

ISBN 978–0–19–928594–5

10 9 8 7 6 5 4 3 2 1

Preface

The title may bring to mind some classic works: *Geometry and the Imagination* by Hilbert and Cohn-Vossen, *The Psychology of Invention in the Mathematical Field* by Hadamard, and *Mathematical Discovery* by Polya. But the associations are misleading. This book is not a mathematical text *à la* Hilbert and Cohn-Vossen, not a psychological investigation *à la* Hadamard, nor a How-To manual *à la* Polya. It is a work of epistemology. But unlike almost all other writing in epistemology of mathematics, it is constrained by results of research in cognitive science and mathematics education. So the book has interdisciplinary roots. And I have tried to make it accessible and interesting to people regardless of departmental boundaries: philosophers, cognitive scientists, educationalists, and historians with a serious interest in mathematical thinking, and to mathematicians inclined to reflect on how they think when at work. That has not been easy. There is always the danger that specialists in one field will find those parts of the book most concerned with other fields difficult or boring, and those parts concerned with their own field irritatingly superficial or unbalanced.

The bulk of the work for this book was done during two years as a British Academy Research Reader. That was undoubtedly the most fruitful period of my intellectual life so far, and I am enduringly grateful to the British Academy for that most precious gift: an uninterrupted period for learning and research, long enough to make real progress. I hope that this book will be judged a worthy outcome.

The work of four people not personally involved in any way with this book has been crucial to my research. Christopher Peacocke's penetrating work on concepts, in particular his landmark volume *A Study of Concepts*, opened avenues that had previously seemed impassable to me. Stephen Palmer's work on visual shape perception was itself an education for me, and it provided resources which, combined with Chris Peacocke's insights, enabled me to explain and substantiate a Kantian claim: the possibility of synthetic *a priori* knowledge in geometry. Stephen Kosslyn's work on visual imagery and its uses, and the detailed picture of an integrated system of vision and visual imagery in his book *Image and Brain*, was invaluable in helping me unlock several mysteries: How is it possible to see the general in the particular? What is the nature of a mental number line? How can there be a visual component to our grasp of abstract structures? The fourth person is Brian Butterworth. I have had the benefit and pleasure of Brian as a teacher and friend. His mastery of the field of numerical cognition, revealed in his book *The Mathematical Brain*, and his own contributions to that field, have provided the foundations of my understanding of basic arithmetical thinking, and inform my chapters on mental number lines and visual aspects of calculation.

Only one of those four is a philosopher; none is a specialist in philosophy of mathematics. But I do have debts to some fellow specialists. Paolo Mancosu's work on explanation in mathematics, philosophically careful and mathematically well informed, has been an inspiration to me. It is always encouraging to find that you are thinking along the same lines as someone whose work you admire, and I have had the extra support of Paolo's personal encouragement and friendship. Michael Resnik's work on mathematical structuralism, in particular his excellent book *Mathematics as a Science of Patterns*, has fed into my thinking; and in my chapter on cognition of structure I discuss and extend one of his proposals. Mike too has been supportive, and I am immensely grateful to both.

I would like to thank the referees, one of whom is the historian of mathematics Jeremy Gray. Both gave me a fund of useful comments that helped me improve the book. "Philosophy is a wordy subject," wrote Jeremy, "and it is easy to get lost." Too true. In revising I tried to make the prose crisper and the signposting clearer, and I have thrown out whole sections to increase cohesion. But there remain parts that are difficult because the problems dealt with are difficult, where greater brevity would have meant less clarity. I would like to thank Jesse Norman for the benefit of extended discussion about diagrammatic thinking when he was my doctoral student and for comments on this book. Progenitors of several chapters were published in refereed journals or presented as talks, so I am grateful to numerous people for comments and discussion at earlier stages. When I have remembered who made a particular point that I mention in these pages, I have acknowledged that person in the notes if not in the text. I thank Peter Momtchiloff of Oxford University Press for steering this book through the publication process. And I thank my wife Frances for her love and support, and for helping me give myself permission to leave the desk occasionally and join her for a walk on the still beautiful Hampstead Heath.

M.G.
London 2005

Contents

1. Introduction — 1
2. Simple Shapes: Vision and Concepts — 12
3. Basic Geometrical Knowledge — 35
4. Geometrical Discovery by Visualizing — 50
5. Diagrams in Geometric Proofs — 71
6. Mental Number Lines — 90
7. Visual Aspects of Calculation — 121
8. General Theorems From Specific Images — 137
9. Visual Thinking in Basic Analysis — 163
10. Symbol Manipulation — 191
11. Cognition of Structure — 214
12. Mathematical Thinking: Algebraic *v.* Geometric? — 240

Bibliography — 269

Index — 285

1
Introduction

What this book is about

Visual thinking, thinking that involves visual imagination or visual perception of external diagrams, is widespread in mathematics, across levels, across subjects, and across kinds of mathematical activity.[1] In support of this claim there is the evidence of school textbooks and research in mathematics education,[2] undergraduate experience in trying to solve problems and understand new mathematics, teachers' experience in presenting mathematics in class or on the page, research experience in attempting to make discoveries and to construct proofs or definitions, and scientists' experience in attempting to devise mathematical methods and models of natural phenomena. The importance of visual thinking in mathematics, then, is not news. But a time-honoured view, still prevalent, is that the utility of visual thinking in mathematics is only psychological, not epistemological. Visual images or diagrams may illustrate cases of a definition, thereby giving us a more vivid grasp of its applications; they may help us understand the description of a mathematical situation or the steps in some reasoning given sentence by sentence; they may suggest a proposition for investigation or an idea for a proof. Thus visual representations have a facilitating role. But that is all, on the prevalent view. They cannot be a resource for discovery, justification, proof, or any other way of adding epistemic value to our mathematical capital—or so it is held. The chief aim of this work is to put that view to the test. I will try to show how, why, and to what extent it is mistaken.

Epistemological questions are evaluative: Can a visual way of acquiring a mathematical belief justify our believing it? Can a visually acquired mathematical belief be knowledge in the absence of independent non-visual grounds? If so, what level of confidence would be rational? In such cases, what we are evaluating is a cognitive state acquired in a certain way.

The investigation therefore demands some account of the way in which the relevant cognitive state is acquired. This entails looking into ways the visual system operates to produce this or that mathematical belief state. What kind of visual representations are used? How are they deployed? What is the nature of the causal route from visual experience (of sight or imagination) to mathematical belief? These are questions of empirical fact, and to answer them we need to draw on findings of the cognitive sciences. The evidence suggests that there are no simple uniform answers to these questions. In many cases, perhaps in most cases, we are not yet able to answer these questions definitively. What we can do is to suggest possibilities that are consistent with what the cognitive sciences have established, and then scrutinize states acquired in those possible ways for epistemological assessment. That will be my course here. Thus another aim of the book, but one essential to fulfilling the chief epistemological aim, is to give some account of ways in which vision and visualization may produce mathematical beliefs, given what is known today. This of course means that some of the book will be devoted to cognitive matters, especially the cognitive science of visual perception and visual imagination.

Proof is not the main concern. Only one chapter is devoted to the question whether visual thinking can be a non-superfluous part of the thinking involved in proving or following a proof, and that is in the context of geometry. The main but not exclusive focus is on discovery. I do not mean merely hitting upon the right answer or some true idea. To discover a truth one must come to believe it independently and in an epistemically acceptable way. Coming to believe something *independently* is just coming to see it by one's own lights, as we say, rather than by reading it or being told. A way of acquiring a belief is *epistemically acceptable* if it is reliable as a way of getting beliefs and it involves no violation of epistemic rationality in the circumstances. Absence of irrationality in getting a belief does not entail that the believer has a justification. Nor is possession of a justification a further condition of discovery: discovery without a justification is at least a conceptual possibility. In mathematics, I believe, that possibility is quite often realized. The correspondence of mathematicians (and one's own experience) suggest that in many cases proofs post-date discoveries. Discovery is sometimes understood to require priority, especially in history of mathematics or science. But that is not how I will be using the word: I will assume that one can discover something that is already known by others. The mathematical focus will not be the discoveries of research mathematicians but the discoveries within reach of practically all of us.[3] Accordingly, most of the examples are drawn from basic mathematics, with occasional forays into university level mathematics in later chapters. The questions of epistemology and cognitive psychology addressed in this book need to be answered as much for basic mathematics as for advanced, and if we have not found the answers for easy and familiar mathematics, we can hardly expect to find them for mathematics that is more complicated and abstruse. So I have deliberately kept the mathematics simple.

The historical context

The reliability of visual thinking in mathematics, especially in analysis (that part of mathematics underlying integral and differential calculus), came under heavy suspicion in the nineteenth century.[4] The main reason was that our visual expectations in mathematics, known collectively as geometrical or spatial intuition, quite often turned out to be utterly misleading, particularly about what happens "at the limit" of an infinite process.

A prominent case is the existence of counter-examples to the "intuitive" belief that a continuous function must have a derivative everywhere except at isolated points. If a continuous function does not have a derivative at a certain point, it has no tangent at that point, and so its curve forms a sharp peak or a sharp valley at that point. We have a strong visual inclination to think that there must be parts of the curve either side of the sharp peak or valley that are smooth enough to have tangents. But continuous functions were discovered that at *no* point have a derivative, known as continuous nowhere-differentiable functions. A function of this kind has no tangent at any point, and so its "curve" would have to consist of sharp peaks or sharp valleys at every point, a possibility that defies visual imagination.

Here is the basic idea of a simplified version of Bolzano's example of a continuous but nowhere-differentiable function.[5] The function is the limit of a sequence of functions, all defined on the interval from 0 to 1 inclusive. The first function in the sequence is the identity $f_1(x) = x$, whose graph is the ascending diagonal from the origin to $\langle 1, 1 \rangle$. The second function f_2 replaces the diagonal by three straight line segments, from the origin to the point $\langle 1/3, 3/4 \rangle$, thence to $\langle 2/3, 1/4 \rangle$, thence to $\langle 1, 1 \rangle$. Thus the first piece of f_2 rises to three-quarters of the height of the diagonal and goes one-third of the way along, the second piece of f_2 falls to one-quarter of the height of the diagonal and goes another third of the way along and the third rises to the total height and goes the final third of the way along. This is illustrated in Figure 1.1(a). The third function f_3 replaces each piece of f_2 by three new pieces; each ascending piece is replaced in the same way that f_1 was replaced by f_2, and the middle descending piece is replaced by three pieces, each going a third of its horizontal span, the first falling three-quarters of its vertical span, the second rising by a half of its vertical span and the last falling to its endpoint. This is illustrated in Figure 1.1(b). In general f_{n+1} is constructed from f_n by replacing each piece of f_n by three new pieces in the same way. It is possible to prove that these functions approximate ever more closely to a single function which is continuous but at no point has a derivative.

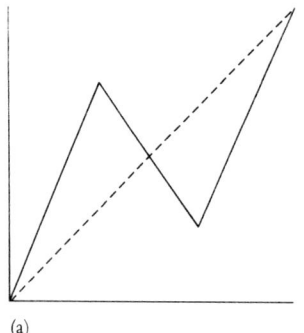 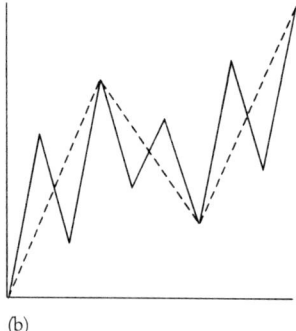

(a)   (b)

Figure 1.1

A related example reinforced suspicion. In 1890 Peano showed that it is possible to define a curve that completely fills a two-dimensional region, where a curve is any set of points onto which the unit interval of real numbers [0, 1] can be continuously mapped. This appears to be impossible, as a curve with endpoints would seem to be a figure with length but not area. Since Peano's original example of a space-filling curve, many others have been found by considering the limit of an infinite sequence of ordinary curves. Hilbert explained the geometric idea underlying one way of generating a sequence of curves whose "limit" curve fills a square. Figure 1.2 illustrates the first three steps of the generation of the Hilbert curve.[6]

Such cases seemed to show not merely that we are prone to make mistakes when thinking visually—that is also true of thinking that is non-visual and

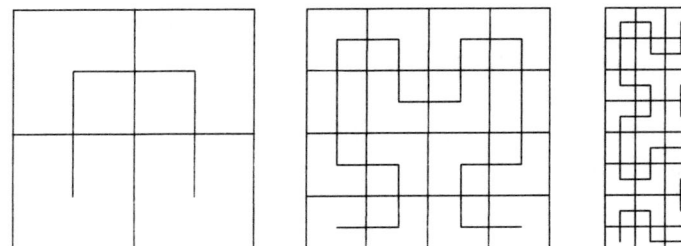

Figure 1.2

numerical—but also that visual understanding actually conflicts with the truths of analysis. Unlike geometrical fallacies such as the isosceles triangle fallacy, in these cases there is no easily correctable visual assumption that accords with our visual comprehension of space and spatial objects.[7] The subject matter seemed to be beyond the visualizable realm. Whether that is in fact so is the topic of a later chapter. But that was certainly common opinion in the late nineteenth century and throughout the twentieth century.

The suspicion of visual thinking was one factor motivating the late nineteenth-century drive to give mathematics a rigorous reformulation in terms of numbers and classes, without reference to anything spatial. That project seemed to be reaching a successful conclusion when, at the end of the nineteenth century, certain paradoxes about classes were discovered which had no obvious solution. This initiated a period of intense research into the logical foundations of mathematics, and it promoted an even stricter attitude about acceptable thinking in mathematics. But there was no consensus as to the correct logical restraints, and major camps in twentieth-century philosophy of mathematics (predicativism, finitism, and intuitionism) were centred on what needed to be done in order to place mathematics on sure foundations. The traditional concern with the way an individual can come to know specific mathematical truths, a concern found in the writings of Plato, Kant, and Mill among others, was entirely abandoned. The focus was on whole bodies of mathematics, such as number theory, analysis, and set theory, regarded as axiomatic systems. It was simply assumed that mathematical knowledge would have to be a matter of proof, that is, deduction from the axioms; the only question, then, was how the axioms and inference rules of the relevant axiomatic systems could be justified. Thus the epistemology of individual discovery simply dropped off the agenda. So did any concern with actual thinking in mathematics.

Three kinds of response to the problem of justifying axiomatic systems for classical mathematics came to the fore. One view of mathematical axiomatic systems is that they are simply bodies of logico-linguistic conventions. Each such corpus of conventions can only be justified pragmatically, in terms of convenience and fruitfulness; there is neither need nor possibility of establishing the axioms true and the rules valid. This view, espoused by Carnap in mid-twentieth century, was persuasively opposed by Quine.[8] The conventionalist view is implausible: surely the axioms of arithmetic given by Dedekind and Peano are true. Our justification for thinking so, according to Quine, is that they constitute an indispensable part of our predictively most successful total science. Similarly for other mathematical systems, Quine argued. Our justification for believing them is empirical, deriving ultimately from observations of events correctly predicted by a combination of theories, including those theories we deem mathematical. This is the second kind of response, holistic empiricism. What is striking about holistic empiricism is that it treats mathematics and natural science indistinguishably; it entails that even professional mathematicians must await the verdicts of empirical science before they can justifiably assert the truth of their mathematical beliefs. Many people have found this hard to swallow.[9]

But if one finds neither conventionalism nor holistic empiricism plausible, what alternative is there? The alternative on offer was a recidivist appeal to mathematical intuition, put forward in brave isolation by Gödel.[10] What neither conventionalism nor holistic empiricism can account for is the *cogency* of the axioms of set theory, "the fact that the axioms force themselves upon us as being true".[11] Mathematical intuition is called on to explain this fact. But what is intuition? It is like sense perception, but differs from it in that the data of intuition are not sensations and are not caused by actions of things on our sense organs. It does not follow, Gödel argued, that the data of intuition are subjective, as Kant had asserted. "Rather they, too, may represent an aspect of objective reality, but as opposed to the sensations, their presence in us may be due to another kind of relationship between us and reality."[12] The problem for this view is that no one has any idea what this relationship is or what cognitive faculties are involved.

Thus the dominant problem of twentieth-century mathematical epistemology, the justification of axiom systems for bodies of established mathematics, received no satisfactory solution. One reaction to this situation has been to shift attention away from axiomatic systems. In recent years there has been a growing interest in mathematical practice among philosophers.[13] However, this interest is still rather narrowly focused. When philosophers of mathematics consider mathematical activity, as opposed to bodies of established mathematics, they tend to think of the research activity of professional mathematicians in proving theorems. What other activities are there? A preliminary list might include discovering truths, explaining them, formulating axioms or definitions, constructing problem-solving techniques, constructing methods for applications, and devising symbol systems.[14] For each of these there are really three different kinds of activity. For a discovery there is the primary activity involved in *making* it; but there is also the activity of *presenting* it, by means of talks, demonstrations, journal articles, or books; and there is the activity of *taking in* the presentation by audience or readers. The trio of making, presenting, and taking in obtains also for other kinds of endeavour on the list. The makers are primarily research mathematicians, pure and applied, though not exclusively. Physicists and in an earlier age, amateur mathematicians, play a prominent part. The presenters, by contrast, include not only mathematicians but also teachers. The takers-in include not only mathematicians and teachers, but also apprentices, students, and schoolchildren. So mathematical activity thus broadly conceived is something that most of us indulge in at some time.

Why not, then, reopen the investigation of earlier thinkers from Plato to Kant into the nature and epistemology of an individual's basic mathematical beliefs and abilities? Why not look at every kind of thinking in mathematics, starting with the simplest, in order to understand its nature and assess its epistemic standing? One does not need to go very far along this road in order to notice the omnipresence of visual thinking in mathematics. Further exploration reveals a diversity of kinds of visual operation in mathematical thinking, and these can be used in different ways. There is no reason to assume that a uniform epistemic evaluation will fit all cases; on the contrary, we already know that there is an epistemic difference between rule-guided formal symbol manipulation, regarded by Hilbert as the most secure kind of thinking in mathematics,[15] and the images of "what happens in the limit" that we use in analysis. Thus an extensive field for research, still largely unexplored, opens up before us. What makes this investigation apposite now is the maturing of the cognitive sciences, especially in the realm of vision and visualization. This is an advantage we have over philosophers of earlier times. If we use it properly, the hope of progress is not unrealistic.

Plan of this book

The book can be thought of as falling into three parts, though the division has no significance from an epistemological point of view.

> The first part, Chapters 2 to 5, is about geometry.
> The second part, Chapters 6 to 8, is about arithmetic.
> The third part, Chapters 8 to 12, is not restricted by mathematical area. It covers further topics, including what there is about analysis and algebra.

I regret to say that there is nothing about topology. The first part, about geometry, needs to be read first, and the chapters of that part need to be read in order. Otherwise the dependency of later chapters on earlier ones is fairly slight, though the second part of Chapter 11 depends on Chapter 6. I will now describe each of the three parts in a bit more detail.

The opening chapter of the geometric part, Chapter 2, is devoted to the cognitive resources needed for some basic geometry. Using that material, I try to show in Chapter 3 how we might come by a simple geometrical belief without inferring it from other beliefs, in a way that is consonant with the apparent obviousness and cogency of the acquired belief. This is intended to be a model for an explanation of beliefs that, cognitively speaking, constitute ultimate premisses. I also try to show that a belief so acquired can be knowledge, and I respond to objections on that score. Chapter 4 attempts to show how, using basic beliefs, one can go on to make a geometrical discovery by visual means in a non-empirical manner. I focus on a simple example, in order to illustrate the general possibility of what Kant would call synthetic *a priori* judgements in geometry; and I try to show how such a judgement can be knowledge. It is commonly asserted that diagrams have no non-redundant role in a proof, even in a geometric proof. "A theorem is only proved when the proof is completely independent of the diagram", wrote Hilbert in lecture notes on geometry.[16] I examine the case for this assertion in Chapter 5, and find it wanting.

The part on integer arithmetic opens with Chapter 6 on the nature and use of mental number lines. This chapter looks to see what cognitive science has to tell us about mental number lines, and I argue that they constitute a resource that is more basic and more important than is commonly appreciated. Chapter 7 is about calculation. I try to dispel the idea that visual thinking is peripheral, especially in the calculations which furnish us with our knowledge of simple numerical equations. In light of what is known about youngsters' calculation, I consider the epistemology of such knowledge, concluding that both Kant and Mill were partly right and partly wrong. Chapter 8 examines the nature of the thinking in a couple of "pebble arguments" for general theorems of number theory. The central question is whether this kind of path to a general theorem constitutes a genuine way of discovering it. With qualifications the conclusion I reach is positive.

The final part looks beyond the very elementary mathematics treated in earlier chapters. Chapter 9 considers the vexed question of visual thinking in analysis. Is visualizing in analysis just a facilitator? Or can it have a non-redundant role in discovering truths of analysis? I argue that its role in discovery is necessarily highly restricted; but several other important functions are fulfilled by visual means in analysis. Chapter 10 explores the nature and uses of visual thinking with symbols in mathematics. I examine visual symbolic thinking, which is more varied than one might expect, to see how it can contribute to discovery, security, illumination, and generality. I also look at the roles of symbolic thinking in certain algebraic examples. Chapter 11 investigates the nature of our grasp of structures by means of visual experience. Structures, which are properties of structured sets of things, are abstract. Does this mean that we can only know a structure under some theoretical description, as the structure of models of this or that theory? Or can there be some more direct mode of comprehension, one that involves visuo-spatial experience? I present the case for the latter possibility. This chapter depends on Chapter 6. In Chapter 12, the final chapter, I try to get some perspective on mathematical thinking as a whole. What kinds of thinking does it comprise? In particular, I scrutinize the common appeal to a distinction between algebraic and geometric thinking. By looking at a number of examples I show that what underlies this appeal is not a division but two poles of something more like a spectrum; and that any division of mathematical thinking into just two kinds is bound to be misleading. Mathematical thinking is richly diverse, and a taxonomy of mathematical thinking that has any worth will have to reflect this fact.

Notes

1. For what I have in mind by "mathematical activity" see Giaquinto (2005b).
2. This should be understood to include relevant developmental studies on preschoolers such as those of Bryant (1974). Some of those studies, plus more recent ones, are reported in Bryant and Squire (2001); they reveal the relevance of spatial thinking in early mathematical development.
3. This is one respect in which my project is quite different from that of Hadamard (1945). Another is that Hadamard was only concerned with the psychology, not the epistemology, of discovery in mathematics.
4. For a short account of this and other factors in the drive to rid analysis of anything spatial and recast it solely in terms of sets and sequences of numbers see Giaquinto (2002: pt. I, ch. 1).
5. Bolzano's example was found in his unpublished work of 1834. In 1860 Cellérier gave another example, though it was not published until much later. Weierstrass gave another example in 1872 that was published in 1875. For a mathematician's response to such findings see Hahn (1933).
6. Hilbert constructed his example in 1891, a year after Peano. For the full story of space-filling curves with mathematical proofs, see Sagan (1994).
7. The isosceles triangle fallacy is an argument that every triangle is isosceles. The pivotal fallacy is the tacit assumption that for any triangle the angle bisector at a vertex V and the perpendicular bisector of the side opposite V meet inside the triangle. It is easy to show visually that there are counter-examples.
8. Carnap (1947); Quine (1960). Russell and Wittgenstein also gave expression to conventionalist views, though Russell oscillated and Wittgenstein was tentative.
9. But holistic empiricism has been ably developed and defended in Resnik (1997).
10. Despite its name, intuitionism, by which I mean the school led by Brouwer, did not appeal to intuition to justify axiom systems for classical mathematics. Classical mathematics is in error, according to intuitionism. Objects and truths of mathematics are mentally constructed; they are not discovered by intuition or anything else. See van Atten (2002: ch. 1).

11. Gödel (1964). At that time many including Quine did *not* find the axioms of set theory compelling. Since then, however, Gödel's attitude to the axioms has spread. So the point holds today, and it held even then with respect to the axioms for arithmetic.
12. Gödel (1964).
13. Tappenden (2001) gives a concise and insightful overview.
14. For discussion and illustration of various mathematical activities, see Giaquinto (2005*b*).
15. Hilbert (1925).
16. Hilbert (1894).

2
Simple Shapes: Vision and Concepts

Plato famously presented a visual way of discovering a simple fact of geometry: if a diagonal of one square is a side of another square, this other square has twice the area of the first.[1] Generalizing from his discussion of this case, Plato gave a tentative account of how geometrical knowledge is possible. That account has been much disputed, but a satisfactory alternative is hard to come by. So the question raised by Plato—how is pure geometrical knowledge possible?—is still very much alive today. Over the next three chapters I give my answer. In any case of geometrical discovery there must be some starting points. Sometimes these starting points are previous geometrical discoveries; sometimes they are truths that just seem obvious and are not the result of prior reasoning. These ultimate starting points may be thought of as cognitive axioms. But in order to be starting points for genuine discovery, it is not enough that these truths strike us as obvious. We must actually know them to be true. They must constitute basic geometrical knowledge. So an initial challenge is this: How can we acquire *basic* geometrical knowledge?

Here are the bare bones of the answer that I will try to substantiate in what follows. Our initial geometrical concepts of basic shapes depend on the way we perceive those shapes. In having geometrical concepts for shapes, we have certain general belief-forming dispositions. These dispositions can be triggered by experiences of seeing or visual imagining, and when that happens we acquire geometrical beliefs. The beliefs acquired in this way constitute knowledge, in fact synthetic *a priori* knowledge, provided that the belief-forming dispositions are reliable. This is the skeleton of my answer to the question of basic geometrical knowledge. The aim of this chapter is to provide the material for fleshing out my proposal in the next chapter.

The material that I will need to draw on comprises, first, certain aspects of two-dimensional shape perception and then an account of perceptual and geometrical concepts for squares.

Aspects of Two-Dimensional Shape Perception

From the pattern of light falling on the retina of each eye the brain must somehow deliver perception of a shape. The light stimulates various kinds of retinal cells. These cells project back to a region in the occipital lobe, the visual cortex,[2] in a way that initially preserves the arrangement of cells in the retina. Thence processing divides into two streams: the dorsal stream, via cells projecting forward up into the parietal lobe—this appears to supply spatial information for motor behaviour—and the ventral stream, via cells projecting forward down into the temporal lobe—this serves object recognition and perhaps all conscious visual perception.[3]

Recordings of single cells in the primary visual cortex show that they respond selectively to highly specific image properties. One class of cells, for example, responds maximally to edges at a particular orientation, while others respond maximally to blobs or bars at that orientation, yet others to edges, blobs, or bars at other orientations.[4] Generalizing from what has been said so far, perceptual processing in the primary visual cortex and beyond appears to consist in analysis of replicas of the retinal image by means of specialist neurons that supply information to other specialist neurons for further analysis, and so on, finally delivering object perception. But this picture has been found wanting in several ways. It suggests (a) that the processing in visual perception produces representations only of features of the retinal image, (b) that processing involves no input except that deriving from retinal stimulation (the processing is all "bottom-up" and never "top-down"), and (c) that there are no intermediate perceptions which affect the end result. There is strong evidence to the contrary, and illustrations are not hard to come by. My present concern is with (a). Here is a well-known counter-example: in seeing the Kanizsa figure as a triangle (Figure 2.1, left), the visual system responds as though detecting straight edges from vertex to vertex when no such edges occur in the retinal image; similarly the visual system responds to the array of horizontal line segments (Figure 2.1, right) as though detecting an inner pair of vertical edges.[5] Visually perceiving

14 SIMPLE SHAPES: VISION AND CONCEPTS

edges or borders of surfaces, then, does not necessarily involve seeing lines that mark those borders.

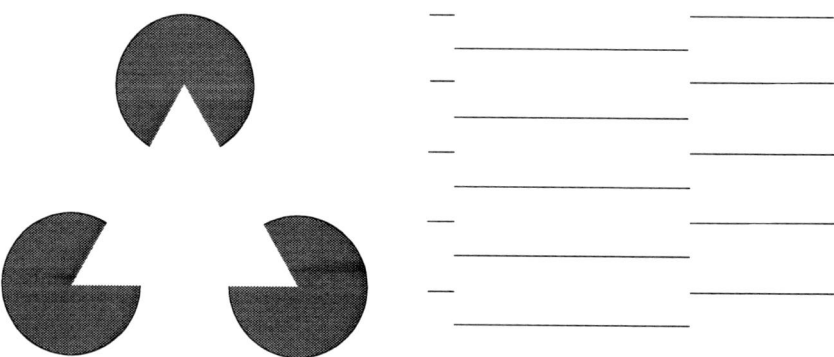

Figure 2.1

Perceptual borders can be constructed from contrasts in the luminance of adjacent regions, or from lines, or by the more indirect means involved in perceiving illusory borders in figure 2.1.[6] Visual object recognition often requires parsing the scene into bordered segments.[7] The visual system typically extracts an arrangement of bordered segments, which is then the input for further operations. The surfaces of a scene very rarely form a jigsaw of perfectly fitting parts in a plane; usually some surfaces will be in front of and partly occluding others. In reaching a representation of a set of surfaces the visual system must determine which are in front of which.[8] An important factor in determining which surfaces are in front and which behind is stereopsis. Another factor, one that operates in the perception of two-dimensional diagrams and pictures, is border assignment. One determinant of border assignment is thought to be the T junctions. When one part of the top of a T junction is common to two regions, the border containing the top of the T is assigned to the region that does not contain the stem of the T, and that region is seen as a surface that occludes part of a surface to which the adjoining region belongs (Figure 2.2).

In some circumstances these dispositions of the visual system can be overridden. Although we can see the middle diagram of Figure 2.2 as a representation of a rectangle occluding part of a surface that might be a full octagon, we can also perceive it as a rectangle adjoining a semi-octagon in the same plane. The relevant contextual fact is that the figures can be seen

SIMPLE SHAPES: VISION AND CONCEPTS 15

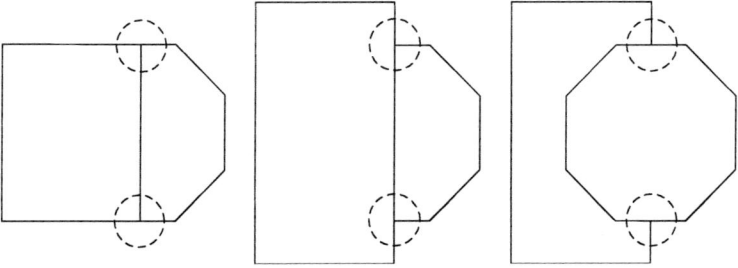

Figure 2.2

as belonging to a single plane surface, the page. So in two-dimensional shape perception, which will be our concern, common borders are assigned to both regions, regardless of cues that would be used in depth perception.

Orientation and reference systems

Perception of an object or figure can be radically affected by its orientation. A well-known example first introduced and discussed by Ernst Mach is the square-diamond.[9] A square with a base perceived as horizontal will be perceived as a square and not as a diamond; but a square perceived as standing vertically on one of its corners will be perceived as a diamond, not a square (Figure 2.3). Irvin Rock drew attention to other examples, such as the difficulty in recognizing familiar faces in photographs presented upside down and failure to notice that a figure is the outline of one's country when it is presented at 90° from its familiar north-up orientation.[10]

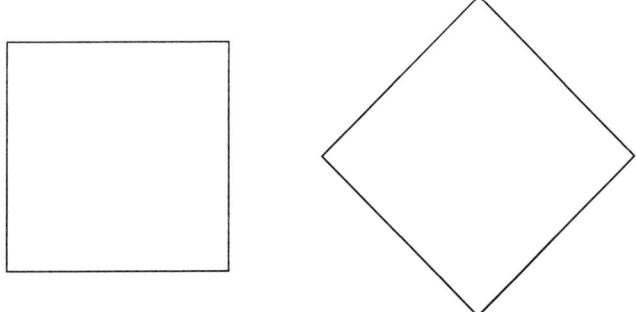

Figure 2.3

Orientation is relative to a reference system. A *reference system* (RS) is a pair of orthogonal axes, one of which has an assigned "up" direction.[11] A reference system can be based on features of the perceived object, on the perceiver's retina, head, or torso, on the edges of a page (if the object is a diagram), or on the environment (horizon plus gravity). Rock stressed the importance of specifying which reference system is operative when making claims about the effects of orientation on perception. A change of reference system is liable to alter perceptual outcomes. Suppose you are looking at a symbol on a page from the side. Switch from a head-based to a page-based reference system and what was perceived as a capital sigma (Σ) may come to be perceived as a capital em (M) (Figure 2.4). Perceptual processing prefers some reference systems to others. View a square with its sides at 45° to floor and ceiling and, as mentioned earlier, it will appear as a diamond; tilt your head 45° so that the figure has sides that are horizontal or vertical with respect to the retinal axes and it will *still* be perceived as a diamond. This is because, in the absence of additional factors, the visual system prefers environmental axes to retinal axes.

Figure 2.4

By thinking of the content of a visual representation as a set of feature descriptions we can make sense of the dependency of perception on orientation. Descriptions in the description set of a visual representation use a selected reference system. When a square is perceived as a diamond the description set will include the information that the object is symmetrical about the vertical (up-down) axis with one vertex at its top, another at its bottom, and one vertex out to each side. That description will not be in the description set for perception of a figure as a square; that will include instead the information that the object is symmetrical about the vertical axis with one horizontal edge at its top, another at its bottom, and one vertical[12] edge out to each side. The description sets are different; hence the perceptions have different contents. Similarly, the description set for Σ contains "centred horizontal line at top", which is not in the description set for M. The description set for M contains "vertical lines at each side", which is not in the description set for Σ. In fact the descriptions of both sets taken together are mutually contradictory[13] unless two systems of axes are used.[14]

A couple of warnings about description sets may be helpful. First, the descriptions in a description set are not the perceiver's commentary and they are not expressed in a natural language. They are simply representations of visual features, encoded in a format that has neural realization. A description set is a set of associated feature representations, and it has much the same role as a category pattern in the category pattern activation subsystem postulated by Kosslyn.[15] Secondly, perceivers do not have conscious access to the description sets of their perceptual representations. When we experience a representational change in viewing a constant figure (duck to rabbit; upright Σ to fallen M), there is a change of description sets. But we rarely know just what changes of description are involved.

When we see a figure in an unusual orientation, such as the letter R on its side, how do we recognize it as an R? To answer this we need to distinguish between mere perception and perceptual recognition. Perception involves generating a set of descriptions of what is perceived; recognition involves this and the additional step of finding a best match between the generated description set and a stored description set for the conventional appearance of the figure. In the case under consideration, the letter R on its side, page-based and head-based axes coincide and have the same up and down directions, and that will be the preferred reference system, initially at least. The conventional top of the figure, however, will not be perceived as top, since relative to that preferred reference system it is off to one side rather than vertically above; and for the parallel reason its conventional bottom will not be perceived as bottom. Matching needs the conventional top and bottom of the figure to be top and bottom with respect to the up and down directions of the preferred reference system.

This can be achieved by selecting a different reference system, one that assigns up and down directions to an axis that is horizontal with respect to the page and head. Changing the axis is something we can do at will. If, for example, subjects are told that the top of a figure is 45° clockwise (or "North-East"), that will affect subjects' perceptions just as if the preferred

18 SIMPLE SHAPES: VISION AND CONCEPTS

up-down directions of their visual system had changed.[16] An alternative is to visualize the figure rotating about its centre in the plane of the page until conventional top and bottom of the visualized figure are top and bottom with respect to the vertical axis of the reference system. Roger Shepard and his co-workers found striking evidence that we can match two three-dimensional figures presented on screen by visualizing rotations.[17] They also found evidence that we can recognize alphanumeric characters in unusual orientations by visualizing rotations.[18] Recognition by matching in either of these ways, changing the reference system or visualizing a rotation, can lead to discovery. If we perceive a square-diamond only as a square because of its orientation on the page, we may come to believe that it is also a diamond through the visual experience resulting from visualizing the figure rotate by 45° or selecting as the vertical of the reference system an axis at 45° to the page-based vertical.

Intrinsic axes and frame effects

To recap briefly, the visual system usually prefers an environmental reference system to an egocentric reference system, e.g. gravitational over retinal axes, when these do not coincide; and both may be overridden by consciously directed attention, even when they do coincide. In fact they may be overridden *without* directed attention when the figure viewed has a strong intrinsic axis. For example, an isosceles triangle with large equal angles and a narrow third angle will be perceived as having the narrow vertex as its top and the short side opposite the narrow vertex as its base, even if the narrow vertex is way off pointing up with respect to environmental, egocentric, and page-based axes. The bisector of the narrow vertex is the intrinsic axis to which the visual system is drawn, and so one naturally perceives the triangle in Figure 2.5 as tilting and pointing in the "North-East" direction, while the accompanying figure, lacking a strong intrinsic axis, need not be seen that way.

When is a line through two points on the perimeter of a figure a strong intrinsic axis? Let the part of a line falling within the boundary of the figure be called its internal segment. One proposal is that if the internal segment of one such line is significantly longer than all the others, such as the internal segment of the major axis of an obvious ellipse, that line will be the figure's strong intrinsic axis.[19] There is some evidence that length is an important factor in recovering the descriptions of a perceived shape.[20] But

SIMPLE SHAPES: VISION AND CONCEPTS   19

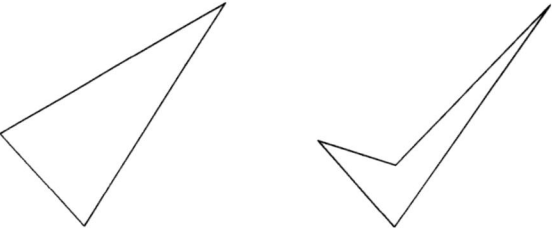

Figure 2.5

the results of experiments on frame effects led Stephen Palmer to conclude that reflection symmetry outweighs length as a determinant of the intrinsic axis used by the visual system.[21] If a figure has more than one pair of orthogonal axes of symmetry, which pair of axes the visual system uses as reference axes depends on surrounding features of the scene. Palmer and his colleagues showed that equilateral triangles can be perceived as pointing in one of three directions and the selected direction depends on contextual features.[22] Compare the central equilateral triangles in the three-triangle arrays in Figure 2.6. Although all the triangles have the same orientation with respect to page and retina, the triangles in the left array are likely to be seen initially as pointing in the 11 o'clock direction while those in the right array are likely to be seen initially as pointing in the 3 o'clock direction. This can be explained in terms of the coincidence of axes of symmetry. In the left-hand array the 11 o'clock symmetry axes coincide, whereas in the right-hand array their 3 o'clock symmetry axes coincide.

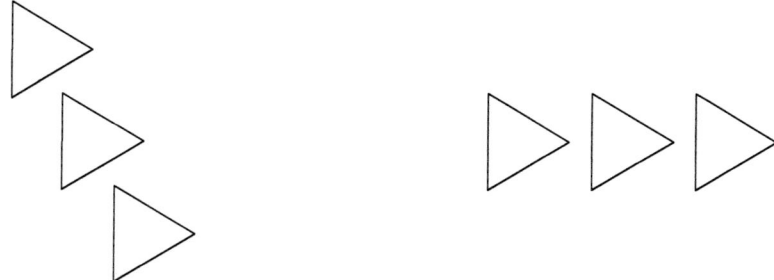

Figure 2.6

The role of reflection symmetry

Other frame effects can also be explained in terms of axes of reflection symmetry. Earlier I mentioned Mach's observation that a square with

20   SIMPLE SHAPES: VISION AND CONCEPTS

sides at 45° to the page and to the retinal axes is seen not as a square but as a diamond, unlike a square whose sides are horizontal or vertical with respect to those axes. But the Mach phenomenon can be offset by additional configurations, such as other squares or a rectangular frame, as in Figure 2.7.[23]

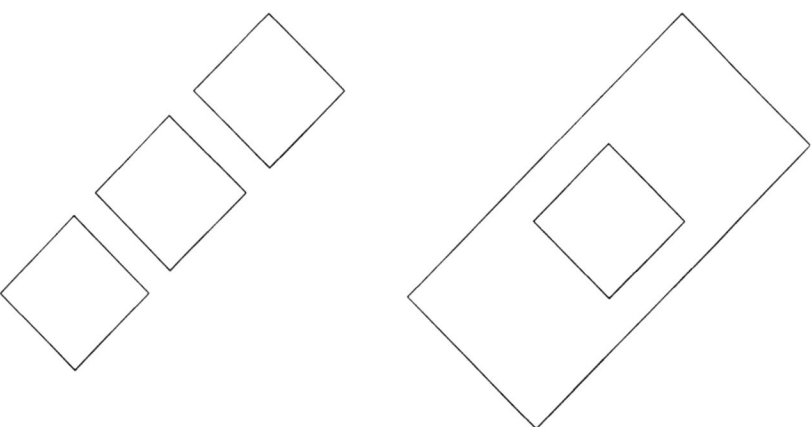

Figure 2.7

Although the central squares are in the diamond orientation with respect to page and retina, one sees them as squares. Palmer explains this by the fact that the bisector of the diamond's upper right and lower left sides is the diamond's only symmetry axis that on the left (in Figure 2.7) coincides with a symmetry axis of the accompanying diamonds and on the right with a symmetry axis of the surrounding rectangle. One sees a square as a diamond rather than a square just when the visual system uses an axis through opposite *vertices* as the main up-and-down axis; one sees it as a square rather than a diamond just when the visual system uses an axis through opposite *sides* as the main up-and-down axis. When a symmetry axis of one figure coincides with a symmetry axis of one or more surrounding figures, the visual system is more likely to use that axis as the main up-down axis for feature descriptions. It is as though there were an augmentation rule for the salience of a symmetry axis: the salience of a symmetry axis of a figure increases when the figure is accompanied by another figure symmetrical about the same axis. Figure 2.8, adapted from one of Palmer's figures,[24] illustrates this for the configuration on the right in Figure 2.7. Symmetry axes are shown for the rectangle, the diamond, and then the two combined.

SIMPLE SHAPES: VISION AND CONCEPTS   21

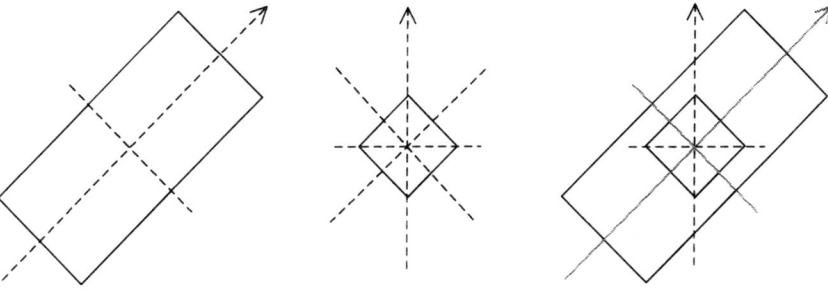

Figure 2.8

Before investigating description sets for these shapes, an apparent circularity in this account must be removed. To perceive a figure as having a certain reflection symmetry, the visual system must first select the relevant axis as an axis of possible reflection symmetry. But in the account given above, in order to select an axis for generating descriptions, the visual system must first determine the figure's reflection symmetries. This problem disappears if symmetry perception involves two processes: a fast but rough test of reflection symmetry in all orientations simultaneously; then, if one or more axes of symmetry are detected, a more precise evaluation of symmetry about one or more of these axes in turn.[25] The initial selection of reference system axes for generating descriptions depends only on the rough and rapid process of symmetry detection, while perceiving the figure as having a certain reflection symmetry (so that that symmetry is in the figure's description set) depends on the second more precise evaluation process.[26]

Once axes are selected, reflection symmetries may have a further effect on which feature descriptions are generated, hence, on which features are perceived. This can be illustrated by examining a square in the normal orientation and in the diamond orientation. Since the reflection symmetries about the selected axes are perceived, features entailed by those symmetries may also be perceived. Look first at the square in normal orientation, on the left in Figure 2.9. It is perceived as symmetrical about its vertical and horizontal axes. But it would not look symmetrical about the vertical axis unless its upper angles looked equal and its lower angles looked equal. It would not look symmetrical about the horizontal unless the angles on the left looked equal and the angles on the right looked equal. So perceiving these symmetries entails perceiving every pair of adjacent angles as equal. Perceiving these symmetries simultaneously, one perceives all the angles as equal.

22 SIMPLE SHAPES: VISION AND CONCEPTS

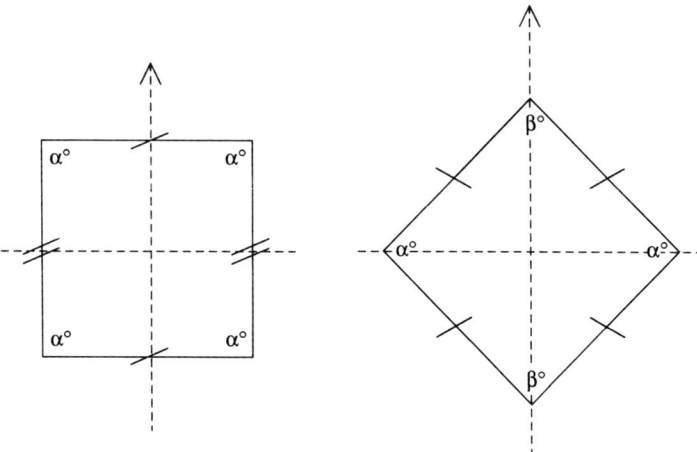

Figure 2.9

Compare this with the perception of angles of the diamond on the right in Figure 2.9. In that case perceiving the symmetries entails perceiving *opposite* angles as equal, not on perceiving adjacent angles as equal. If all pairs of adjacent angles are equal, all angles are equal; but all pairs of opposite angles may be equal without all angles being equal. If we pressed the left and right vertices closer together so that the angle $\alpha°$ is greater than the angle $\beta°$ of the top and bottom vertices, the figure would still be perceived as a diamond. So perceiving a figure as a square entails that all its angles are perceived as equal, whereas perceiving it as a diamond does not.

We can thus explain the Mach phenomenon in terms of the selection of reference systems with axes parallel to page edges and with the same "up" direction as the page. Though the square and the equiangular diamond have the same shape and size, their different orientations with respect to those reference systems produce different feature descriptions, hence different perceptual contents.

How does perception of the symmetries about the vertical and horizontal axes relate to the perception of the *sides* of the figures? It is the reverse of the relation between perceiving the symmetries and the angles. When seeing the figure as a diamond, simultaneously perceiving the symmetries about the vertical and horizontal axes entails perceiving all its sides as equal. When seeing it as a square, simultaneously perceiving those symmetries entails perceiving opposite sides as equal, not on perceiving all its sides as equal.[27]

But we do perceive all sides of a square as equal and this is what distinguishes seeing a figure as a square from seeing it merely as a rectangle. The visual system can pick up this additional information by means of a secondary set of orthogonal axes, the axes at 45° to the primary pair of axes. Perceiving the symmetries of the square about these diagonal axes involves perceiving adjacent sides as equal, and that is enough to distinguish the square from other rectangles in perception. As there are now two pairs of orthogonal axes involved, the visual system must discriminate between them. Only one of these pairs can be used as the reference system for descriptions. The axis-pair of the reference system is primary, in that it is selected before feature descriptions are generated. A secondary pair of axes can be singled out by reference to the primary pair. In this case the other axes might be described as angle-bisectors of the primary pair.

An alternative possibility is that the visual system uses co-ordinates based on the axes of the reference system with a Euclidean metric. Using this co-ordinate system, a measuring mechanism computes lengths, which serve as inputs to a system that encodes spatial properties based on size, as distinct from a system for coding non-metric spatial relations such as connected/apart, inside/outside, above/below.[28] Using the co-ordinate system we could detect equality of sides, and thereby distinguish squares from other rectangles. But a more economical way of achieving this end is by perceiving the symmetries about the diagonal axes, and so I will assume that this is how the visual system operates.[29]

A category specification for squares

The foregoing provides all the ingredients needed for a category specification for squares. Let V and H be the vertical and horizontal axes of the reference system. Then to perceive a figure as a square it suffices that the visual system detects the following features.

Plane surface region, enclosed by straight edges:
  edges parallel to H, one above and one below; edges parallel to V, one each side.
Symmetrical about V.
Symmetrical about H.
Symmetrical about each axis bisecting angles of V and H.

My suggestion is that this is (or might be) a visual category specification for squares, in the following sense. When, in seeing or visualizing something, the description set for the visual representation contains descriptions of all the features in this category specification, what is seen or visualized is seen or visualized as a square.[30] This is not to say that these are the only features of squares that we perceive. But whenever the visual system detects these features, the figure will be perceived as a square. In short, these features are enough, though they are not all. The fact that this category specification is so economical may help to explain why squares are so basic to our visual thinking. Of course this may not be the category specification for squares that is actually operative in all or any of us; if the visual system uses a category specification for square recognition that is different but not wildly different, the details of my subsequent story would have to be changed. But the approach would remain intact.

Concepts for squares

The capacity to reason about squares is distinct from the capacity to recognize perceptually something as a square.[31] The capacity to reason about squares requires that one has a concept for squares. A concept for some kind of thing is not to be identified with a perceptual category specification for that kind. This is most obvious for functional kinds: the features in one's perceptual category specification for telephones, the features used by the visual system in recognizing something as a phone, are not essentially related to its function; but our concept of phone is essentially functional. But concepts for perceptible properties, in at least some cases, are intimately linked to the corresponding perceptual category specification. How certain concepts for squares relate to the perceptual category specification for squares is the subject of the second half of this chapter. Before turning to that, it will help to outline the general approach to concepts that will be adopted here.[32]

A concept, as that term is used here, is a constituent of a thought, and a thought is the content of a possible mental state which may be correct or incorrect and which has inferential relations with other such contents.[33] Neither thoughts nor concepts are here taken to be linguistic entities. Though thoughts can often be expressed by uttering a sentence

SIMPLE SHAPES: VISION AND CONCEPTS 25

and concepts can often be expressed by words or phrases in a sentence, thoughts are not taken to be sentence meanings, and concepts are not taken to be lexical meanings.[34]

Why assume that thoughts have constituents? We need to account for inferential relations between thoughts. Consider the following inferences.

1. Mice are smaller than cats. Cats are smaller than cows. Therefore mice are smaller than cows.
2. Tom was an uncle. Therefore Tom was a brother or brother-in-law.

To explain the validity of the first inference we would naturally look to what is expressed by "smaller than", as that is what is common to all three thoughts. The validity of the second depends on the connection between what is expressed by "uncle", "brother", and "brother-in-law". Why regard the things expressed by these terms as *constituents* of thoughts? Why not just say that they are combined to form a thought, without implying that they are somehow *in* the thought? The reason is that there is some sense in which each of them typically has a position in the thought. For example, the same concepts are combined in the thoughts expressed by "Tom's uncle is a father" and "Tom's father is an uncle", but the thoughts are different, because the concepts are differently ordered.[35]

So concepts, as I am using that word, are constituents of thoughts on which some of their inferential relations depend. Christopher Peacocke has pointed out that we can characterize a concept in terms of these relations. To possess a concept one must be disposed to find certain inferences cogent without supporting reasons.[36] So we can in principle specify a concept in terms of these basic inferences.[37] For example, we can specify a concept for uncles thus:

{**uncle**} is that concept C which one possesses if and only if one is disposed to find inferences of the following forms cogent without supporting reasons:

$C(x)$. Therefore, for some person y, x is a brother or brother in law of a parent of y.

For some person y, x is a brother or brother in law of a parent of y. Therefore x C y.

A natural question about this view of concepts is: How does concept possession get off the ground? Getting the concept {**uncle**} requires that we already possess concepts for brother and parent. This might suggest that to get the rich variety of concepts people usually have, there would have to be a substantial innate stock of concepts, which is hardly plausible. This is not in fact a consequence of the "inferential role" view of concepts, since non-conceptual content can enter into the specification of the conditions for possessing a concept.[38] Types of transition from one content-bearing mental state to another are essential to the individuation of a concept, on this view. But these states may sometimes be experiences with non-conceptual content rather than thoughts. How this might go will be revealed in the discussion of a perceptual concept for squares.

A perceptual concept for squares

Before proceeding to specify a *geometric* concept for squares, I will specify a *perceptual* concept for squares. This is because the initial geometric concept for squares that I am aiming at is a restricted version of a perceptual concept for squares, and is most easily explained if the perceptual concept is introduced first. The actual specification of a perceptual concept will be quite complicated. But the root idea is that the perceptual concept for squares centres on a disposition to judge something square when it appears square and one does not suspect that circumstances are illusiogenic or one's vision is malfunctioning. This kind of account would be circular (hence fail to specify any concept) if it were not possible for something to appear square to a person without that person's deploying the perceptual concept for squares. As a square figure appears square just when its squareness is perceived, we can see that this is possible, by noting the distinction drawn earlier between merely perceiving the squareness of a figure and perceptually recognizing the figure as square. Recognizing a figure as square involves a perceptual experience of it which draws on an antecedently acquired category specification for squares: not only must visual processing generate the descriptions in the category specification for squares, it must also make a best match between the set of generated descriptions and the previously stored category specification. But for merely perceiving the squareness of a figure it will suffice if visual processing generates the descriptions of the category specification for squares. It is not also required that the description set is matched with the descriptions of a stored category specification for squares; in fact no antecedently stored representation of any kind (for squares) needs to be accessed. Thus there is no pressure at all to hold that perceiving the squareness of a figure must involve deploying a concept for squares.[39]

To specify a perceptual concept for squares we use the feature descriptions in the category specification, but allow for imperfections, as we can recognize a figure as a square even if, for example, it is visibly not completely enclosed or its sides are visibly not perfectly straight. The degree of imperfection allowable is not something one can specify; obviously the lines must be sufficiently straight and the figure sufficiently enclosed to generate the feature descriptions "straight line" and "closed figure", and so on. I will use the modifier "n/c" for "nearly or completely" to make this explicit. I will use "V" and "H" as before for the vertical and horizontal axes of the reference system. A further point must be taken into account when giving the perceptual concept. We can apply perceptual concepts to things that we are not perceiving. To cater for this the perceptual concept will have two parts, for the cases in which one is thinking about a perceived item and an unperceived item respectively. Here goes.

The concept {**square**} is the concept C that one possesses if and only if both of the following hold:

(a) When an item *x* is represented in one's perceptual experience as a n/c plane figure n/c enclosed by n/c straight edges, one edge above H and n/c parallel to it, one below H and n/c parallel to it, and one to each side of V and n/c parallel to V, and as n/c symmetrical about V and n/c symmetrical about H, and as n/c symmetrical about each axis bisecting angles of V and H—when *x* is thus represented in perceptual experience and one trusts the experience, one believes without reasons that that item *x* has C. Conversely, when one trusts one's perceptual experience of an item *x*, one believes that *x* has C only if *x* is represented in the experience as a n/c plane figure n/c enclosed by n/c straight edges... etc.

(b) Let "Σ" name the shape that figures appear to have in the experiences described in clause (a). When an item *x* is unperceived one is disposed to find inferences of the following form cogent without supporting reasons:

x has Σ. Therefore *x* has C.
x has C. Therefore *x* has Σ.

Obviously what perceptual concept for squares one actually possesses depends on the feature descriptions actually used by the visual system in perceiving something as square. The set of feature descriptions given earlier is not the only one that is consistent with the data about shape perception. Also, it is at least theoretically possible that different people have different perceptual concepts for squares. So it would be wrong or at least potentially misleading, given the present state of knowledge, to talk of *the* perceptual concept for squares. But there is no impropriety in talking of *the* perceptual concept {**square**}, since that is the concept just specified.

A geometrical concept for squares

The perceptual concept {**square**} is a vague concept; that is, there may be things for which it is indeterminate whether they fall under the concept. This is because there is some indeterminacy in the extensions of perceptible properties described in the visual category specification for squares, such as near or complete straightness and near or complete reflection symmetry. Among things which are clearly square, such as a handkerchief or the surface of a floor tile, we can sometimes see one as a better square than another: edges sharper, straighter, or more nearly equal in length, for example, corners more exactly rectangular, halves more symmetrical,[40] and so on. Sometimes we can see a square, one drawn by hand for instance, as one which could be improved on and we can imagine a change which would result in a better square. It can be part of the content of an experience of those having the concept {**square**} that one square is a better square than the other.

It can also be part of the content of experience that a square is perfect. Since there is a finite limit to the acuity of perceptual experience, there are lower limits on perceptible asymmetry and perceptible deviation from (complete) straightness. Asymmetry about an axis which is so slight that it falls below the limit will be imperceptible; similarly for non-straightness. So for any figure veridically perceived as symmetrical about an axis α, if its asymmetry about α falls below the lower limit of perceptible asymmetry, it will be perceived as maximally symmetrical about α; similarly for straightness. Hence there is a maximum degree to which a bounded plane surface region can be perceived as symmetrical about a given axis, and a maximum degree to which a border can be perceived as straight. When in experiencing something as square these maxima are reached, this fact may be encoded in the description set generated in the perceptual process. In that circumstance the perceived item will be experienced as having perfectly straight sides, and as perfectly symmetrical about vertical and horizontal axes. If in addition the same applies with respect to the other features in the category specification for squares, the item will appear perfectly square.

To be more precise, let us say that an item *appears perfectly square* when it is represented in one's perceptual experience as a perfectly plane figure completely enclosed by perfectly straight edges, one edge above H and perfectly parallel to it, one below H and perfectly parallel to it, and one to each side of V and perfectly parallel to V, and as perfectly symmetrical about V and perfectly symmetrical about H, and as perfectly symmetrical about each axis bisecting angles of V and H.

Just as possession of the *perceptual* concept for squares centres on a disposition to judge something square just when it appears square and one trusts the experience, so possession of an initial *geometrical* concept of squares centres on a disposition to judge something square just when it appears *perfectly* square to one and one trusts the experience. The only difference is that the features that figure in the geometrical concept must be perfect exemplars of their kind. Hence where, in the specification of the perceptual concept, the figure or parts of it are required to be *nearly or completely* this or that (e.g. n/c straight edges), in the specification of the geometrical concept they must be *completely* or *perfectly* this or that. This is the only difference between the perceptual concept and the initial geometrical concept.

There are certainly other concepts expressed by the word "square" that qualify as geometrical concepts. But this or something close to it is probably right for basic geometrical knowledge. It does not seem plausible that we have a geometrical concept for squares that is not similarly linked to a perceptual concept of the kind given earlier, unless it is a concept that depends on theoretical concepts from e.g. real (or complex) analysis, the theory of complete ordered fields, or something similarly abstruse. I assume that geometrical concepts that are modifications of perceptual concepts can be given for other entities of basic geometry and for their basic geometrical properties and relations. In the next chapter I will try to show how having such concepts enables us to get basic geometrical knowledge.

Notes

1. See Plato's dialogue *The Meno* 81e–86c. The translation I favour is Plato (1985).
2. The visual cortex consists of the striate cortex and adjacent areas referred to as the extrastriate cortex.
3. The ventral and dorsal streams have been described as the "what" and "where" streams respectively, following ideas put forward by Ungerleider and Mishkin (1982). More recently Goodale and Milner (1992) have suggested instead that the ventral stream is for conscious perception and the dorsal stream for visual control of skilled motor actions. See also Goodale (1995).
4. The primary visual cortex in humans is the area V1. But initial evidence comes from research with cats and macaque monkeys by Hubel and Wiesel (1962, 1968). Evolutionary and anatomical considerations provide indirect evidence that these findings are true of humans as well, and there is some direct evidence that low-level visual processing is similar in macaques and humans (DeValois et al. 1974).
5. Against (b), Jastrow's famous Duck-Rabbit can be seen as a duck or as a rabbit, and one can switch from one to the other at will without any change of retinal stimulation. Against (c), seeing the Duck-Rabbit figure either way requires seeing the line as the outline of an almost enclosed region (and not just a snaky line) so that there is a perception mediating the final outcome.
6. The expression "illusory border" is customary but misleading, as an illusory border can be the result of a real border in the physical world. What is meant is just that there is no corresponding border in the retinal image.
7. Segmentation, hence object recognition, can be difficult without borders. A striking example is given by Kundel and Nodine (1983). This is presented and discussed by Ullman (1996: 235–42).
8. For a full account see Nakayama et al. (1995).
9. Mach (1897).
10. Rock (1973).
11. If one axis is assigned up-down direction, surely the other is assigned left-right direction? Not necessarily. Reorienting figures by reflection about their vertical axis has little effect on perceived shape (Rock 1973).

This could be because the horizontal axis is not assigned left-right direction—but then codes of features on one side of the vertical axis would have to be bracketed together somehow, to separate them from codes of features on the other side. For further evidence that shape descriptions do not include left-right information see Hinton and Parsons (1981).
12. "Vertical" abbreviates "parallel to the vertical axis".
13. There is no reason in principle why a description set cannot contain contradictory descriptions. Perhaps this occurs when one has a perceptual illusion of "impossible" events or objects, such as the waterfall illusion or the Penrose triangle.
14. This may sometimes happen: a familiar symbol (letter or numeral) on its side can be recognized but still look slightly strange, perhaps because a page-vertical axis is still weakly operative even though the vertical axis of the reference system used for shape recognition is page-horizontal, with up and down directions that coincide with the conventional up and down of the symbol.
15. Kosslyn (1994).
16. Rock and Leaman (1963).
17. Metzler and Shepard (1974).
18. Cooper and Shepard (1973).
19. Marr and Nishihara (1978).
20. Humphreys (1983).
21. Palmer (1990).
22. Palmer (1980). See also Palmer and Bucher (1981).
23. Palmer (1985). Rock (1990) cites Kopferman as the source of the frame effect illustrated on the right in Figure 2.7.
24. Palmer (1985).
25. If the axes of probable symmetry include vertical, horizontal, 45° diagonal axes, and others, that is likely to be the order in which they are evaluated for symmetry. See Goldmeier (1937), Rock and Leaman (1963), and Palmer and Hemenway (1978).
26. This two-process model was proposed by Palmer and Hemenway (1978) to account for response-time data that conflict with the predictions of a model consisting of a single sequential symmetry evaluation procedure proposed by Corballis and Roldan (1975). A two-process model was also used by Bruce and Morgan (1975) to account for differences they

32 SIMPLE SHAPES: VISION AND CONCEPTS

found among small violations of symmetry, including the fact that some are detected much faster than others. Finally, Driver, Baylis, and Rafal (1992) found that figure-ground perception by a patient with unilateral neglect could be affected by reflection symmetry, while the patient was unable to perform tasks requiring explicit perception of reflection symmetry—evidence for two kinds of capacity for detecting reflection symmetry, one that does not require attention and one that does.

27. This is an elaboration of the analysis in Palmer (1983). Evidence that these are the relevant features distinguishing the two shape perceptions comes from the fact that the figures we perceive as squares we view as special kinds of rectangle, not special kinds of rhombus, whereas squares we perceive as diamonds we view as special kinds of rhombus. Palmer cites Leyton, 'A unified theory of cognitive reference', *Proceedings, Fourth Annual Conference of the Cognitive Science Society* (1982).

28. This is motivated by experimental results reported in Kosslyn et al. (1989).

29. Evidence that we do sometimes use two pairs of orthogonal axes is that we can see e.g. a W tilted with respect to the page *as* a tilted W. Seeing it as a W requires perceiving its conventional top and bottom as up and down with respect to a conventional vertical axis, and seeing it as tilted involves perceiving the conventional vertical at an angle to the page-based vertical.

30. How would the description set for perceiving a figure as an equilateral diamond differ from this? It will not contain the feature of sides parallel to the axes V and H of the reference system. A candidate description set for the equilateral diamond is this: Plane surface region, enclosed by straight edges, with vertices on H, one to each side of V, and vertices lying on V, one to each side of H; symmetrical about V; symmetrical about H; symmetrical about each axis bisecting angles of V and H.

31. It is hardly controversial that if one cannot recognize squares because one has lost sight and touch, one has not thereby lost the capacity to reason about squares. But the possibility of a capacity to recognize squares without a capacity to reason about them is contestable. In my view it is possible, because one may lack awareness of the features used by the visual (or tactile) system in recognizing squares (or one may lack

the grade of awareness needed for reasoning); so recognitional ability would not entail a corresponding reasoning ability. I do not know of an actual dissociation of this kind.
32. This is essentially the approach of Peacocke (1992).
33. The term "concept" is also used for word sense, explanatory theory, category representation, prototype (i.e. representation used in typicality judgements), and others. See the introduction of Margolis and Laurence (1999).
34. In general, thoughts and concepts seem too fine to be identified with expression meanings. But I do not insist on this view, if only because the individuation of expression meanings is such a slippery, obscure, and contentious matter.
35. I thank Samuel Guttenplan for pointing out the need to justify the idea that concepts are constituents of thoughts. I accept that what I have said here is only a partial justification, and that the analogy between concept–thought relation and the phrase–sentence relation may turn out to be misleading.
36. This is a slight oversimplification, if inferences are transitions between thoughts. In some cases one needs to consider transitions from content-bearing mental states that are not thoughts. My use of the word "inference" here is intended to include such transitions.
37. See Peacocke (1992). This view of concepts (thought constituents) is minimalist. It is consistent with the approach taken in this paper that minimalism misses out something essential to the nature of concepts. For example, one might hold that part of what constitutes possessing a perceptual concept for squares is that one has a symbol for the perceptual category of squares. In this spirit Giuseppe Longo (personal communication) has suggested that what I am talking about might better be called "proto-concepts".
38. For an introduction to the idea of non-conceptual content, see Crane (1992). To see how the idea fits into a general theory of content, see Peacocke (1992) and (1994). McDowell (1994) attacks the view that experience can have non-conceptual content. For responses and counter-responses see Peacocke (1998), McDowell (1998), and Peacocke (2001).
39. Nor is it required that one deploys a concept for any of the features in the description set for squares, such as straightness or reflection

symmetry, though it is required that the visual system can detect and represent these features.

40. See Palmer and Hemenway (1978) for evidence of our ability to make judgements of approximate reflection symmetry. For a definition of a variable symmetry magnitude see Zabrodsky and Algom (1996). Degree of symmetry (using this definition) was found to correlate fairly well with perceived figural goodness.

3
Basic Geometrical Knowledge

How can we acquire basic geometrical knowledge? By "basic knowledge" I mean knowledge not acquired by inference from something already known or from an external authority, such as a teacher or book. My short answer to the question is that in having geometrical concepts we have certain general belief-forming dispositions that can be triggered by visual experiences; and if that happens in the right circumstances, the beliefs we acquire constitute knowledge. Drawing on the material of the previous chapter, I will try to substantiate this answer by focusing on a way of acquiring the particular geometrical belief that the parts of a square either side of a diagonal are congruent.[1]

From concepts to belief-forming dispositions

Concept possession may bring with it a belief-forming disposition. I will try to show this for the case of someone possessing both the concept {**perfect square**} specified in the previous chapter and a concept for restricted universal quantification. Restricted universal quantification is what is expressed by phrases of the form "All Fs" or "Every F", as in "Every man has his price".[2] The key fact about the concept of restricted universal quantification, which I will denote {**r.u.q**}, is this:

> If one has the concept of restricted universal quantification, one will believe the proposition "Every F has G" when and only when one would find cogent any given inference of the form "x has F, so x has G".

Now suppose that having these concepts, {**perfect square**} and {**r.u.q**}, you perceive a particular surface region x as perfectly square. You can think of its apparent shape demonstratively, as *that* shape. Letting "S" name that shape, your coming to believe of some item that it has S will result in your believing that it is perfectly square; and your coming to believe of some item that it is perfectly square will result in your believing that it has S (as long as you can think of S demonstratively).[3] That is, you will find any given inference of the following forms cogent: "x has S, so x is perfectly square" and "x is perfectly square, so x has S". This follows from the conditions for possessing the concept {**perfect square**}, which are the conditions used to define the concept in the previous chapter. As you also have the concept {**r.u.q**}, you will have the following disposition, which I will denote "PS" for "perfect square":

> (PS) If you were to perceive a figure as perfectly square, you would believe of its apparent shape S that whatever has S is perfectly square, and that whatever is perfectly square has S.

If you have this disposition, merely seeing a figure as perfectly square will produce in you a pair of general beliefs. Although these beliefs are not logical trivialities, they are not empirical truths either, as epistemic rationality does not require that one has evidence for these beliefs; in particular one does not need to inspect a sample of things having S for rationally believing that whatever has S is perfectly square. This then illustrates a rational way of getting a belief as a result of a concept-generated disposition triggered by a visual experience. In a similar way, I claim, we can acquire the geometrical belief that the parts of a square either side of a diagonal are congruent.

Suppose one has a concept for geometrical congruence. If a figure *a* appears to one symmetrical about a line *l* and one trusts the perceptual experience, one will believe that the parts of *a* either side of *l* are congruent. We can further say that if *a* appears to one symmetrical about *l*, regardless of whether one trusts the experience one will believe that if *a* were as it appears (in shape), the parts of *a* either side of *l* would be congruent. With this antecedent condition, it is only the apparent shape of *a* that is relevant: having that shape, the shape that *a* appears to have, suffices for the attributed property. So one has a more general belief, about *any* figure having the apparent shape of *a*, that it has the attributed property. This is the level of generality that we require for geometrical truths. Of course, the attributed property in this case is not congruence of the parts of the figure either side of the very line *l*, because the line *l* is just a line through *a*. What we have in mind, for a figure *x* having the apparent shape of *a*, is a line through *x* that would correspond to *l* through *a* if *a* were as it appears. What is correspondence? A line *k* through *b* *corresponds to* line *l* through *a* if and only if some similarity mapping of *a* onto *b* maps *l* onto *k*.[4] To put all this together, suppose one has concepts for perfect correspondence, similarity, and congruence. Then one will have the following belief-forming disposition, which I will call "C" for congruence.

> (C) If one were to perceive a plane figure *a* as perfectly symmetrical about a line *l*, then (letting "S" name the apparent shape of *a*) one would believe without reasons that for any figure *x* having S and for any line *k* through *x* which would perfectly correspond to *l* through *a* if *a* were as it appears, the parts of *x* either side of *k* are perfectly congruent.

Getting the belief

A feature in the visual category specification for squares is near or complete symmetry about a line that bisects the angle made by the vertical and horizontal axes of the reference system, and such a line is a diagonal of the square. As the degree to which a figure's apparent squareness approaches perfection depends directly on the degree to which its apparent symmetry about diagonals approaches perfection, it follows that if a figure appears perfectly square to one, it appears perfectly symmetrical about its diagonals. So, given a concept for diagonals as well as the concepts that provide one with disposition (C), one will have a disposition that is a special case of (C):

> If one were to perceive a plane figure *a* as perfectly square, then (letting "S" name the apparent shape of *a*) one would believe without reasons that for any figure *x* having S and for any line *k* through *x* which would perfectly correspond to a diagonal of *a* if *a* had S, the parts of *x* either side of *k* are perfectly congruent.

For anyone having this disposition, merely seeing a figure as perfectly square will result in their acquiring the belief mentioned above. This should not be assimilated to making a judgement as a result of an observation. First, getting a belief with a certain thought-content can occur without thinking the thought, whereas making a judgement cannot.[5] Secondly, the experience is not the ground for the belief, for one does not need to take the experience to be veridical. The role of the experience is merely to trigger the disposition.

In this case the belief is about the figure perceived and its apparent shape. Yet the target belief is general. To reach the target belief, other dispositions must operate. Recall the disposition (PS) that comes with possessing the concept {**perfect square**}:

> (PS) If you were to perceive a figure as perfectly square, you would believe of its apparent shape S that whatever has S is perfectly square, and that whatever is perfectly square has S.

Because of this, when one who possesses the concept {**perfect square**} perceives *a* as perfectly square, "having S" and "being perfectly square" become cognitively equivalent for that person, in the sense that they will be treated as standing for the same attribute automatically, without any act of inference. So the joint possession of the disposition (PS) and the special case of (C) entails possession of a disposition that is just like the special case except that talk of having S is replaced by talk of being perfectly square. When that disposition is triggered one gets this belief:

> For any perfect square x and for any line k through x which would correspond to a diagonal of *a* if *a* were perfectly square, the parts of x either side of k are perfectly congruent.

To have this disposition one must have a concept for diagonals. Given a perceptually based geometric concept for diagonals, like the concept {**perfect square**}, one would think of a line through perfect square x which would correspond to a diagonal of *a* if *a* were perfectly square as, simply, a diagonal of x. Thus one would have a disposition just like the special case but with this consequent belief:

> For any perfect square x and for any diagonal k of x, the parts of x either side of k are perfectly congruent.

Spelled out, the disposition is this:

> (C*) If one were to perceive a plane figure *a* as perfectly square, one would believe without reasons that for any perfect square x and for any diagonal k of x, the parts of x either side of k are perfectly congruent.

The point here, the truly remarkable point, is that if the mind is equipped with the appropriate concepts, a visual experience of a particular figure can give rise to a general geometric belief. In short, having appropriate concepts enables one to "see the general in particular". One cannot have those concepts without having a disposition to form a general belief as a result of a certain kind of visual experience. In the example at hand the general belief is the target belief that the parts of a square each side of a diagonal are congruent.

But, one may object, one has no awareness of getting this belief from a particular visual experience. Isn't this a problem for the account? Not at all. In very many cases we are unaware of the cause and occasion of the acquisition of a belief. Having a belief is not a manifest state like a pain state—some of our beliefs we are unaware of having—and the transition from lacking a certain belief to having it may also occur without awareness. Were you aware of acquiring the belief, say, that antelopes and chimpanzees cannot interbreed, at the time of acquisition? One may not get a firm belief all at once; to acquire a firm belief by activation of a belief-forming disposition, activations on several occasions may be needed. But the point is unchanged: there is no anomaly in the fact that we are usually unaware of those occasions.

This answers the question how it is possible to acquire this general geometric belief without inference or external written or spoken testimony. Of course, people may get the belief in different ways, and it is an empirical question whether anyone gets the belief in the way that I have described. What is suggested here is merely one possibility. In one respect it is a rather unlikely possibility. How often do we see something as perfectly square? A closely related possibility is one in which the triggering experience is of the kind described except that the figure is seen as a square but not a perfect square. In this case I suspect that one can acquire the target belief in the same way except that the route goes through visual imagination: in perceiving the figure as a square one is caused to imagine a perfect square.[6] This is possible if, as I believe, possession of the relevant concepts gives rise also to belief-forming dispositions the same as those above except that they are activated by visualizing rather than visual perceiving. This is because, I suggest, what is causally operative in the mechanism underlying the disposition is the activation of the visual category specification for e.g. (perfect) squares; it does not matter whether activation is initiated in searching for a best match with a set of perceptually generated descriptions or in executing an intention to visualize a perfect square.

These suggestions are, as I said, mere possibilities, to be eliminated or modified in the light of future findings. But they are partial answers to the Kantian question "How is it possible to have basic geometrical knowledge?" which respect the role of sensory experience without collapsing geometry into an empirical science. The answers are only partial, because nothing has yet been said to show that acquiring a belief in such a way can be knowledge. Let us turn to that question now.

Is it knowledge?

If one comes to believe in the way described above that the parts of a square each side of a diagonal are congruent, is that belief knowledge? To be knowledge the belief must be true, it must have been acquired in a reliable way, and there must be no violation of epistemic rationality in the way it was acquired and maintained. In my view these three conditions suffice for knowledge. It is at least arguable that there is a further condition: the believer must have a justification for the belief. I will briefly address the question in the light of these four conditions, and then respond to a couple of objections. The first condition, that the belief be true, can be dealt with quickly. The belief we are concerned with is not (or not necessarily) about squares in space as it actually is, but about squares in space as it *would* be, if it were as the mind represents it, that is, in Euclidean space.[7] In Euclidean space the parts of a perfect square either side of each diagonal are perfectly congruent. So the belief is true. This leaves the conditions of reliability, rationality, and justification.

Reliability If one reaches the belief in the way described, the belief state results from the activation of the belief-forming disposition. So the question we have to answer is whether this belief-forming disposition is reliable. What does reliability come to here? The kind of belief-forming dispositions we usually assess for reliability have varying output beliefs for varying inputs satisfying a given condition. If the output belief would be true for any input that satisfied the given condition, the disposition is reliable. If in a significant number of possible cases the output belief would be false, the disposition is unreliable. We might say, for instance, that a disposition to believe what you read in a certain daily newspaper is unreliable, knowing that its journalistic practices are slack and unscrupulous. This poses a problem for us, as this criterion of reliability cannot be applied to a belief-forming disposition that has an unvarying output belief, moreover one which is a mathematical truth, as in the case of C*. For such a disposition no possible input would yield a false output belief, however crazy the disposition. One might, for example, have a disposition to believe Fermat's Last Theorem on reading a certain argument for it with a subtle fallacy.

There is a way around this problem. Consider a disposition to believe a particular true mathematical proposition upon following a certain argument with that proposition as its conclusion. If this disposition is to count as reliable, the argument must at least be sound. If the argument contains a fallacious step the disposition is unreliable. We can make sense of this because the disposition to believe the conclusion upon following the argument is the result of other dispositions, including the disposition to accept inferences of the kind instantiated by the fallacious step. This will be an inferential disposition with outputs that vary for varying inputs. That disposition is one to which we can apply the criterion of reliability given above: if the output belief would be true for any input that satisfied the given condition, the disposition is reliable; if in a significant number of possible cases the output belief would be false, the disposition is unreliable. Now we can say that a belief-forming disposition is reliable if it is reliable by this criterion or if this disposition results from other dispositions all of which are reliable by this criterion.

What this means for the disposition C*, which has an unvarying and necessarily true output belief, is that one must consider its causal basis. One has (C*) as a result of having certain other dispositions, namely the disposition (C) together with the disposition (PS) and dispositions issuing from a concept for diagonals. In the case of (C), the output belief varies with varying instances of the antecedent condition. So we can apply the criterion given above. To put it in slightly altered terms: when the antecedent is realized, the consequent belief corresponding to that realization is true. The disposition (C), recall, is this:

(C) If one were to perceive a plane figure *a* as perfectly symmetrical about a line *l*, then (letting "S" name the apparent shape of *a*) one would believe without reasons that for any figure *x* having S and for any line *k* through *x* which would perfectly correspond to *l* through *a* if *a* were as it appears, the parts of *x* either side of *k* are perfectly congruent.

Different instances of the variable *a* give different antecedent conditions and different consequent beliefs. We can tell that (C) is reliable by inspection. Let *a* be a plane figure one perceives as perfectly symmetrical about an axis *l*. Then the apparent shape of *a* is such that if *a* has it *a* is perfectly symmetrical about *l*. So for any figure *x* having that shape and for any line *k* through *x* which would perfectly correspond to *l* if *a* were as it appears, the parts of *x* either side of *k* would be perfectly congruent. Thus the consequent belief would be true. Thus (C) is reliable. In the same way one can check that the disposition (PS) is reliable. The required dispositions that we have in possessing a geometrical concept for diagonals can also be checked for reliability, and I will assume that they and (PS) are reliable without working through the details. As our possession of (C*) is an immediate result of joint possession of those reliable dispositions, this way of getting the target belief, namely by activation of (C*), is reliable.

Rationality For a belief state to qualify as knowledge, certain rationality constraints must be satisfied. Suppose for instance that having acquired a true belief *b* in a reliable way you become aware of having another belief *c* with as much justification as *b* but which is clearly inconsistent with *b*. In this circumstance believing *b* is epistemically irrational and so cannot count as knowledge.[8] There are other rationality constraints. But there is no good reason to think that it is impossible or even difficult to meet rationality constraints. The contrary thought arises when one assumes rationality constraints that are much too strict. An example is the view that consistency of one's belief set is required for avoiding irrationality. This is too harsh because it overlooks the possibility of arriving at a number of jointly inconsistent beliefs, each with explicit justification, when the inconsistency is extremely difficult to detect. That would make the believer unlucky, not irrational. In the absence of any plausible argument to the contrary, I take it that it is possible, perhaps easy, to get a belief by activation of reliable belief-forming dispositions and keep it, without violations of rationality. Thus one may come to believe the target geometrical truth in a reliable way and keep that belief without irrationality. In such circumstances the belief has an epistemically valuable status. I hold that it is knowledge.

Implicit justification Some people say that a belief is not knowledge unless the believer has a justification for the belief. This is too strong if it is required that the believer is able to express a justification, otherwise young children would have very little of the knowledge that we credit them with. A less implausible version of this doctrine requires only implicit justification. What that comes to is not clear; but if it suffices that the person's beliefs and relevant facts about the person's epistemic situation can be marshalled so as to provide a justifying argument, the requirement can be met in the case under discussion. Here is the justifying argument.

1. x is a perfect square. [Assumption]
2. ∴ For any y perceived as perfectly square, x is in shape as y appears. [1]
3. Figure *a* is perceived as perfectly square. [Perceptual state]
4. Anything perceived as perfectly square appears symmetric about its diagonals. [Category specification for squares]
5. ∴ Figure *a* appears symmetric about its diagonals. [3, 4]
6. ∴ x is symmetrical about its diagonals. [2, 3, 5]
7. ∴ The parts of x either side of a diagonal are congruent. [6, by concepts for symmetry and congruence]
8. If x is a perfect square, the parts of x either side of a diagonal are congruent. [7, discharging the assumption]
9. ∴ The parts of any perfect square either side of a diagonal are congruent. [8, universal generalization]

If implicit justification is required for knowledge, it surely cannot amount to more than has been given in this argument. I conclude that the kind of implicit justification that is available, on top of the satisfaction of the conditions of truth, reliability, and rationality, is enough to make the attribution of knowledge safe.

Objections

The route to belief described earlier does not fall within the ambit of currently recognized means of acquiring knowledge, such as sense perception, deductive inference, acceptable inductive generalization, and inference to the best explanation. Moreover, on some views, a belief arrived at in this way has a character that disqualifies it from knowledge. I will consider the two serious objections stemming from such views.

44   BASIC GEOMETRICAL KNOWLEDGE

First Objection: no a priori *knowledge*   On my account a visual experience causes the belief, but does not play the role of reasons or grounds for the belief, as it is not necessary for the believer to take the experience to be veridical: it is enough that a perceived figure *appears* perfectly square. So the visual experience is used neither as evidence nor as a way of recalling past experiences for service as evidence. Rather, the visual experience serves merely to trigger certain belief-forming dispositions. So this way of acquiring the belief is *a priori*, as it does not involve the use of experience as evidence.

But some people hold that there is no *a priori* knowledge. The most influential argument for this view is that *a priori* knowledge, if there were any, would be epistemologically independent of experience; but no belief whatever is independent of experience, because no belief is immune from empirical overthrow. The argument is Quine's, but it starts from a point made by Duhem, that what constitutes evidence for or against a belief depends on what other beliefs we hold fixed.[9] When we find that our beliefs as a whole conflict with observations, we may reject seemingly non-empirical beliefs in order to bring the totality of our beliefs into line with our observations. An oft-cited example is the overthrow of the Euclidean Parallels Postulate. Hence the acceptability of a belief depends on its belonging to a totality of beliefs that fits well with our experience. Thus the acceptability of any belief depends on experience, and so cannot be an instance of *a priori* knowledge.

The most important point in reply to the Quinean objection is this. The way in which the belief about squares in my account is *a priori* relates to its genesis: no experience is used as evidence in *acquiring* the belief. This is consistent with the possibility of *losing* the belief empirically. So even if Quine were right that all beliefs are vulnerable to empirical overthrow, that would not show that beliefs acquired in the way described above could not be knowledge.[10]

Secondly, the argument for the claim that no belief whatever is immune from empirical overthrow is inconclusive. It is true that some seemingly non-empirical beliefs have been overthrown, such as the belief that the shortest distance between two physical points is a straight line. But there may be others that we cannot rationally reject in order to accommodate observations which conflict with the corpus of our beliefs. In particular, it is not clear that the Parallels Postulate has been or could be overthrown empirically. What can be, and perhaps has been, overthrown is the claim that the Parallels Postulate is true of physical space. This is a claim about actual physical space. But the Parallels Postulate itself may be taken as a claim about a certain kind of possible space, about a way that physical space could be, rather than about the way it actually is.[11] Understood that way, the Parallels Postulate has not been overthrown.

Second objection: no conceptual knowledge In the route to belief under discussion, visual experience merely triggers some belief-forming dispositions. Having these belief-forming dispositions is a direct result of having certain concepts. So anyone who has the relevant concepts is bound to get the belief, given a certain stimulus. Moreover, the truth of the proposition believed seems to be guaranteed by the nature of the concepts involved. In this way one can think of the belief as based on the relevant concepts alone. So, if the belief arrived at this way is knowledge, it can be appropriately called conceptual knowledge.

The second objection, put by Paul Horwich, is that there is no conceptual knowledge.[12] The argument is that there may be nothing in reality answering to a concept (no reference or semantic value), in which case a general thought that issues from the concept will not be true. So in order to know that it is true, one must know that the concept has a reference, and that knowledge must have an evidential basis independent of the concept. As an example, consider Priestley's concept of phlogiston. That involves a number of inference forms, e.g. "x is combustible; $\therefore$ x contains *phlogiston*". But nothing simultaneously satisfies all those inference forms; that is, whatever real thing (substance kind) we take as the reference of "phlogiston", not all of those inference forms will be truth preserving. Although the belief that all combustible matter contains phlogiston issues from the concept, that could not be known without knowing that there really is some kind of substance answering to Priestley's concept of phlogiston. Parallel to this, we face a challenge to the reliability of the way in which the belief about squares was reached. We would have to know that there really is a property answering to the concept {**perfect square**}, and we cannot get that knowledge from merely having the concept.

The central claim on which this objection rests is that in order to know a positive general truth we must know that its constituent concepts each have a reference (or semantic value).[13] While this is plausible for concepts introduced explicitly in postulating the existence of something, it is not plausible for other concepts. For example, at a fairly early stage in language development one comes to know that if a statement of the form "*S* and *T*" is true, so is the statement *S*. For this item of knowledge to be true, the concept for conjunction here expressed by "and" must have a reference (or semantic value), which in this case is the truth function for conjunction.[14] But we surely do not come to know that the concept for conjunction has this or any other reference until much later, if at all, when we start thinking about semantics. So the claim on which the objection is based, namely that we cannot know a truth without knowing that each of its constituent concepts have a reference, is too strong. It is reasonable only where a certain class of concepts is involved, including those introduced explicitly in postulating something. We have no reason to think that basic geometrical concepts belong to this class. On the contrary, the concept {**perfect square**} arises naturally from perceptual experience.

Not that a concept which arises naturally is bound to have a semantic value. A natural concept may turn out to be incoherent, as the naïve concepts of "set" and "true" have done. But the significance of this is only that conceptual beliefs are generally defeasible, in the sense that they may (as far as believers can tell) be overthrown by future discovery of conceptual incoherence. It does not mean that conceptual beliefs cannot be knowledge. The situation has a parallel with empirical beliefs: though they are liable to overthrow as a result of future evidence, they can still be items of knowledge. So the objection does not pose a real threat to the knowledge claim made in this chapter.

Summary

How can we acquire basic geometrical knowledge? In answer to this I have presented one possibility for a belief about squares, in the hope that it would serve as a guide in similar cases. In the previous chapter I gave an account of perceiving squareness in which visual detection of reflection symmetries is crucial. This is built into a stored category specification for squares. A perceptual concept for squares uses that category specification, and a basic geometrical concept for squares is obtained by slight restriction of the perceptual concept. This chapter pointed out that possessing these concepts (and others) entails having certain general belief-forming dispositions, which can be triggered by activating the stored category specification for squares either in seeing something as square or by visualizing a square. When a visual experience thus triggers the relevant belief-forming disposition, the experience does not have the role of evidence for the resulting belief. Yet the belief acquired this way can be knowledge: the mode of acquisition is reliable, there need be no violation of epistemic rationality, and the believer has an implicit justification for the belief. Moreover, the only serious objections that I know about can be met.

One final point. This manner of acquiring the belief is non-empirical, because the role of experience is not to provide evidence. At the same time, some visual experience is essential for activating the relevant belief-forming disposition; and it is clear that this way of reaching the belief does not involve unpacking definitions, conceptual analysis, or logical deduction. Hence it must count as non-analytic. Given that "non-analytic and non-empirical" translates as "synthetic *a priori*", we have arrived at a view that is at least close to Kant's often dismissed view that there can be synthetic *a priori* knowledge.[15]

Notes

1. Two figures are *congruent* if and only if they have the same shape and size.
2. Restricted universal quantification is expressed by other phrases too. Examples are phrases of the form "each F" and "any F".
3. The resulting beliefs may be tacit. That is, you may believe that whatever has S is perfectly square without thinking the thought, just as you have believed that none of your grandmothers' grandmothers were elephants, without thinking it before now.
4. A similarity mapping is a shape-preserving transformation, such as uniform expansion or contraction, rotation, translation, or any composition of these. (We include the null transformation among similarity mappings.) Let *a* and *b* be similar, i.e. figures with the same shape. Imagine *a* contracting or expanding uniformly until it forms a figure *a′* the same size as *b*; then imagine *a′* moving so as to coincide with *b*. Any such similarity mapping maps each line through *a* onto a line through *b*.
5. Of course, in normal circumstances in which one has a belief with a certain thought-content, when the question of its correctness is

48 BASIC GEOMETRICAL KNOWLEDGE

explicitly raised in a way that one understands, one will think the thought. But the point here is that seeing a figure as a perfect square may not occasion the thought; it is guaranteed only to produce the belief.

6. This would not appear to be a square separated from (and competing with) the seen square; it might instead be a representation whose activation is involved in recognizing the perceived figure as a square. See Kosslyn (1994: ch. 5).
7. These are the squares that the believer has in mind, but she need not think of this range of squares explicitly, that is under some description such as "squares in Euclidean space". It is of course an empirical matter whether the human mind represents space as Euclidean, as I believe. I do not know of systematic empirical investigations of this question.
8. If there are degrees of belief, this statement should be modified. In the situation described one's degree of belief in b should not exceed one's degree of belief in c, and the sum of those degrees must not exceed the maximal degree.
9. Quine (1960); Duhem (1914).
10. Robert Audi was the first to make this point, I believe.
11. What is this possible way that space could be? I suggest that it is the way that space would be if it were as the mind represents it. Like Kant, I suspect that the Euclidean axioms are a partial articulation of the inbuilt mental representation of space. Unlike Kant, I regard actual physical space as external to and independent of the mind and its representations.
12. See Horwich (1998: ch. 6), and Horwich (2000). Horwich has since refined and elaborated his views. See Horwich (2005: ch. 6).
13. Negative existential truths, e.g. that there is no phlogiston, are not under consideration here.
14. This is the function that takes any ordered pair of propositions to a proposition that is true if both members of the pair are true, and false if at least one member of the pair is false.
15. See Kant (1781–7: Introduction V.1 B14; B16; A25/B39–41). There is of course room for differences of interpretation with regard to Kant's use of the terms "synthetisch" and "*a priori*". Kant claimed explicitly that *all* mathematical judgements are synthetic *a priori*. I take it that he was restricting his attention to true judgements. Even so, I think the

claim is too strong on any plausible interpretation of the key terms. The dominant view in recent times is that no mathematical knowledge is synthetic *a priori*. See Quine (1960) and Kitcher (1984). My view is that, for an epistemically relevant and Kant-like interpretation of the key terms, this claim too is false.

4
Geometrical Discovery by Visualizing

The previous chapter sketched a possible way of acquiring basic geometrical beliefs, and argued that beliefs acquired that way can be knowledge. This chapter is concerned with ways of getting new geometrical knowledge from prior geometrical knowledge. Visual imagination seems to play an important role in extending geometrical knowledge. The central question of this chapter is whether visualizing can be a means of geometrical discovery, and if so, what kind of epistemic role visualizing could play in discovery. Since my aim is to illustrate and argue for a possibility, namely, that visualizing *can* be a means of geometrical discovery, I will proceed in this chapter as in the previous one by focusing on one very simple geometrical example.

Before presenting the example, let me say briefly what I mean by "discovery". As I am using the expression, discovering a truth has three components. First, there is the independence requirement, which is just that one comes to believe the proposition concerned by one's own lights, without reading it or being told. Secondly, there is the requirement that one comes to believe it in a reliable way. Finally, there is the requirement that one's coming to believe it involves no violation of epistemic rationality (given one's pre-existing epistemic state). In short, discovering a truth is coming to believe it in an independent, reliable, and rational way. This specification departs from normal usage in one way that is unimportant for present concerns: it allows that one can discover something that has already been discovered by someone else. This apart, normal usage is respected. In particular the oft-noted distinction between discovery and justification is respected: discovering something does not require that one can justify one's belief in the truth concerned (by proving it or in some other way);

nor is it required that the thinking by which one discovers it constitutes or contains an implicit proof of it. So we have no need to consider the question whether what one visualizes can be re-presented as a proof or part of a proof.[1] The central question is one of reliability. This will be addressed in the discussion of the example that will serve as the focus of this chapter.

The example[2]

Imagine a square. Each of its four sides has a midpoint. Now visualize the square whose corner-points coincide with these four midpoints. If you visualize the original square with a horizontal base, the new square should seem to be tilted, standing on one of its corners, "like a diamond" some people say. Figure 4.1 illustrates this. Clearly, the original square is bigger than the tilted square contained within it. How much bigger? By means of visual imagination plus some simple reasoning one can find the answer very quickly.

By visualizing this figure, it should be clear that the original square is composed precisely of the tilted square plus four corner triangles, each side of the tilted square being the base of a corner triangle. One can now visualize the corner triangles folding over, with creases along the sides of the tilted square. Many people conclude that the corner triangles can be arranged to cover the tilted inner square exactly, without any gap or overlap. If you are in doubt, imagine the original square with lines running between midpoints of *opposite* sides, dividing the square into square quarters, its quadrants. The sides of the tilted inner square should seem to be diagonals of the quadrants.

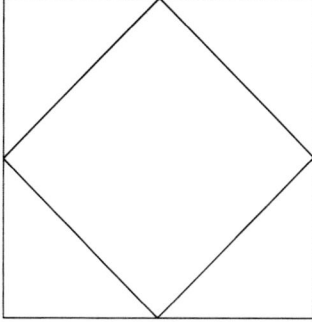

Figure 4.1

Assuming that this leads you to the belief that the corner triangles can be arranged to cover the inner square exactly, you will infer that the area of the original square is twice the size of the tilted inner square. You might reason, for example, as follows. The corner triangles can be arranged to cover the inner square exactly; so the total area of the corner triangles equals the area of the inner square; the area of the original square equals that of the inner square plus the total area of the corner triangles; so the area of the original square equals twice the area of the inner square.

Here is a way of coming to believe a true general proposition about squares in Euclidean space,[3] the proposition that any square c has twice the area of the square whose vertices are midpoints of c's sides. You may have known this already; you may have acquired this belief by having followed a proof of it from certain other beliefs, or by being told, or in some other way. But this should not prevent you from seeing that a person *could have* acquired this belief by visualizing in the way suggested. The route to belief described above is mixed. Part was valid verbal reasoning; part was the act of visualizing which led to one of the premisses of the verbal reasoning. This premiss is the following true belief, that I dub "B":

> [B] If c_i ("the inner square") is the square whose vertices are midpoints of the sides of a square c ("the original square"), then the parts of c beyond c_i ("the corner triangles") can be arranged to fit exactly into c_i, without overlap or gap, without change of size or shape.

The rest of this chapter will focus on the way of arriving at belief B suggested above. As a description of a way of arriving at the belief, what has been said so far is insufficient, and in order to substantiate my positive conclusion I will have to be more specific about the cognitive process that may be involved in getting the belief. Before doing that, however, it will be helpful to consider two possible ways in which the visualizing described above might be used to arrive at the belief. In both of these cases the role of visualizing is to furnish experiential evidence. I will argue that those empirical routes to the belief are not ways of discovering it.

An inference from sense experience?

Suppose that someone not already having belief B acquired it by visualizing in the manner suggested above. Could this have been a genuine discovery? To answer this, we must get a clearer view of the role of the visualizing when belief B is acquired this way. Suppose you are asked how many windows there are in the front of your parents' house, when away from the house. To answer this you might visualize the house front. This would be a way of drawing on visual experience as evidence for the answer. An obvious hypothesis is that the visualizing described above is like this, a way of drawing on experience as evidence bearing on B. In that case the way of acquiring belief B described above is an inference from sense experience. In this section I will argue that, though it may be inference from sense experience, it does not have to be, and that when it is, B is not thereby discovered.

Visualizing the corner triangles of the original square folding over onto the tilted inner square does not seem to involve any inference from sense experience, still less to constitute an inference. But this could be illusory. Suppose you are in a second-hand furniture store looking for a desk to go in the room where you study; you see an attractive desk, somewhat larger than you had been looking for. Would it fit? At this point you visualize the desk in various positions in the room to discover the possibilities. In visualizing thus you surely would be drawing on your experience of the room and its current furniture as evidence for the judgements you make about whether the desk will fit.

Is it not the same when through visualizing one concludes that the four corner triangles of a square would fit exactly onto the tilted inner square without overlap? Having seen a few Christmases come and go, your attempts to wrap gifts and fold paper provide you with experiences which might be drawn upon as evidence for the belief B. When you visualize the corner triangles folding over onto the tilted inner square, are you not generalizing from past experiences of this kind which have been laid down in memory?[4] If so, you would be using these past experiences as inductive evidence, and the visualizing would be a way of drawing on these experiences as a basis for the geometrical conclusion.

Several considerations indicate that it does not have to be so. Here is one. A belief acquired and sustained by inductive inference alone is accompanied by the feeling that it could turn out to be wrong. Consider one's judgement that the large desk one is admiring would fit in one's study along with its other furniture once your old desk is removed. Though you may be pretty confident about it, you will probably accept that when it comes to the physical test you might turn out to be mistaken. This sense of fallibility is even more marked for inductively based beliefs that are general. Even beliefs we are very confident about, if inductively based, will be accompanied by the view that future experience might falsify it. Consider our belief that all snow is naturally white. In fact there may be theoretical backing for the idea that all snow has the same colour, so that all snow is white if any is. But let us ignore this and suppose that our belief that all snow is white was reached solely by induction[5] from experience of snow. We surely feel that it is possible, though unlikely, that we come across decisive counter-evidence in the future, such as areas of red snow.[6] Even with the theoretical backing we feel that the belief might turn out to be false, as the theory may be found wanting and there may be two types of snow. In contrast to this, it is typical that when proposition B is arrived at by visualizing in the way suggested, the believer does not feel that there might turn out to be a counter-example. Our attitude to our paper-folding failures is instructive. Sometimes, when we fail to get the folded corners to fit exactly, there are visible circumstances that account for the failure. For example, the line of one of the folds is not at the correct angle with respect to the sides, or it was not at the right perpendicular distance from the corner-point, or the sheet was not rectangular to begin with. On other occasions when we fail, there are no visible circumstances to account for the failure. But we do not find such would-be counter-examples a threat to our geometrical beliefs, belief B in particular. Instead we suppose that visible failure is the effect of one or more invisibly small imperfections of the kind already mentioned. Typically one feels that a counter-example is impossible.[7] The presence of this feeling is a clear indication that the belief was reached by some means that did not involve inferring it from sense experience.

A second consideration is this. If the act of visualizing were a way of drawing on sense experience as evidence, that experience would have to include fairly extensive confirmation of proposition B, otherwise it would not produce *any* conviction in B, let alone strong conviction. But when you consider the relevant experience, does it not seem a little meagre? How many times have you managed to fold the triangular corners of a piece of paper that seems perfectly square, so that they appear to fit flush over the tilted inner square? A completely successful performance is probably very rare. Perhaps, though, the evidence is more indirect. Maybe it is relevant that you have managed to fold two adjacent corners of an apparently rectangular sheet so that the corners appear to fit flush. But how many times have you done this? The point is that one's successes, those paper-folding experiences which one might have taken as confirming cases, may not be very numerous and may even be outweighed by the failures, and yet by visualizing one may acquire a firm belief in proposition B. This makes it quite implausible that the visualizing must be a way of inferring proposition B from one's paper-folding experiences.

Our attitude to paper-folding successes parallels our attitude to failures. When our physical experiments appear to provide confirming instances of our geometrical beliefs, we suppose that in fact they do not; we suppose that our successful paper-folding attempts merely approximate the geometrical possibilities. For example, we take it that the edges of a sheet of paper are rough, that paper-folds do not crease along perfectly straight lines, that the surface area of the paper is not preserved with the folding, and so on. This we believe holds generally: having fuzzy and inconstant surfaces, physical objects cannot instantiate a theory of perfectly smooth shapes and perfectly rigid transformations.

This attitude towards our paper-folding experiences, namely, that they could not provide inductive evidence for a geometrical proposition like B, need not stop a person from coming to believe proposition B by visualizing in the way suggested. Nor, if the belief was acquired that way, need it be undermined by later acquiring this attitude to one's paper-folding experiences. This provides a third obstacle to the idea that in arriving at B by visualizing one had to be drawing on past experience as evidence for B. For if the role of visualizing could only be to marshal experience as supporting evidence, the thought that one's experience could not warrant the conclusion very probably *would* obstruct the process of arriving at the belief by visualizing; or if one had already arrived at the belief this way, this thought would very probably undermine the belief.[8]

To summarize: one may arrive at belief B by visualizing in the suggested manner, when (a) one feels that a future counter-example is not an epistemic possibility, (b) the putative evidence of sense experience is meagre at best, and (c) one believes that the putative evidence is of a kind which could not warrant belief in proposition B; but if one arrives at belief B by visualizing in the manner suggested under circumstances (a), (b), and (c), it is extremely unlikely if not impossible that the process is a way of inferring B from sense experience. So we can reasonably conclude that arriving at B by visualizing in the manner suggested does not have to involve an inference from sense experience.

Let me make two disclaimers. (1) I am not saying that one could acquire belief B by visualizing without having had some sense experience of physical objects. On the contrary, I think it very likely that such experience has at least two roles in generating a geometrical belief by visualization, both quite distinct from providing evidence for the belief. First, we may need sense experience in conjunction with some innate mental propensities in order to form basic geometrical concepts. Secondly, memories of visual experiences may provide the components on which the mind operates in producing experiences of visual imagination. On this account, sense experience does enter into the causal prehistory of the belief, not as evidence but as raw material from which the mind forms our geometrical concepts and our visualizing capacities. (2) I am not saying that it is never the case that in acquiring a geometrical belief by visualizing one draws on sense experience as evidence. Perhaps a child would deploy visual memories of drawings of triangles to answer the question whether there is a triangle with an internal angle greater than a right angle. I am not even saying that in acquiring belief B by visualizing in the manner suggested, one could not be drawing on sense experience as evidence.

However, if one were drawing on sense experience as evidence in acquiring belief B by visualizing, the result would not be a discovery. This is because B is a claim of *perfect* congruence. It is the proposition that the corner triangles of a square can be arranged to fit *exactly* into its tilted inner square, whereas we have good reason to believe that there could be apparently triangular physical pieces which fit inexactly but well enough to seem exact to the senses. Thus even if there were genuine physical instantiations of geometrical theorems such as B, and even if we find what seems to be an instantiation of B, we could not reliably infer that it is an instantiation of B. Hence the reliability requirement for discovery would not be met. For this reason, if arriving at B by visualizing in the manner suggested can be a way of discovering a truth, it must be possible that the visualizing has a function other than to provide experiential evidence.

An inner experiment?

On an alternative account, the process does involve drawing on experience as evidence, but the experience used is the visualizing experience itself rather than past perceptual experiences of paper-folding and the like. The idea we want to consider is that the experience of visualizing serves as direct and immediate evidence for a judgement about what is visualized, just as an experience of seeing can be immediate evidence for a judgement about what is seen. The visualizing is not simply seeing what is internal to the mind, since what is represented in the experience of visualizing is produced by an *intention* to visualize such-and-such, whereas what is represented in an experience of seeing is not dependent in this way on the perceiver's intentions. Accordingly, visualizing may be thought of as a kind of internal experiment, a process that constitutes both performing the experiment and observing its outcome.

This idea entails that in visualizing the corner triangles of the original square folding over onto the inner square, one is observing (some feature of) that very experience of visualizing. This may seem quite puzzling. How can the experience of *observing* the visualizing experience differ from merely *having* that experience? If it does not differ, what does observing the experience consist in, over and above merely having it? Yet people clearly do sometimes observe features of their own visual experience. Suppose, for example, that you are having an eye test. The optometrist asks you "Which seems clearer, the cross on the green background or the cross on the red background?" or "Do you see two horizontal lines or one?" It is understood that you are not being asked to make judgements about nearby objects. You are being asked to observe and report on features of your visual experience, something you can probably do without too much trouble. In other circumstances and in other sense modalities we sometimes observe features of our own experience: we can observe the shade of a visual after-image or the pattern of variation in the intensity of a pain. In all these cases of inner observation there *is* something over and above merely having the experience: there is the sense of directing one's attention and noticing something as a result, both of which add something to the feel of the experience.

Now what about visualizing in the way suggested earlier? If that is not a way of inferring proposition B from past sense experience, must it be a case of observing something in the experience of visualizing, thence inferring proposition B? Must it be an inner experiment? The examples above suggest not. For the difference in feel that one gets from concentrating one's attention and noticing something in the experience is often missing.

When you visualized the corner triangles folding over onto the inner square, did you *scrutinize* the end-state of the folding, for example, and as a result *notice* that the inner square was covered exactly? The phenomenology of the experience may be relevantly unlike the cases in which one does observe one's own experience, as when undergoing an eye test.

A second difficulty for the view that the visualizing must be, if not an inference from sense experience, an inner experiment, arises from one's certainty in the truth of B: by visualizing in the manner suggested we may come not merely to believe B but to feel certain of it. (A reminder: B says that the corner triangles can be arranged to fit the inner square exactly.) Would we feel certain if the belief depended on observing precisely the fleeting experience of visual imagining? Of course people do become certain of things on the flimsiest of grounds; but in these cases they do not maintain their certainty on reflection, unless their attachment to the belief is enforced by some psychological disposition that induces epistemic irrationality.

We are aware of the fallibility of visual observation of the physical world; optical illusions are common. But there is some temptation to think that we are infallible observers of our own experience, whether this is the experience of seeing or visualizing. However, the difficulty of accurately observing our own visual experience, a difficulty known to older people from eye tests and to others learning to draw or paint what they see as they see it, brings home to us the weakness of our capacities for internal observation. Added to this is the elusiveness of experiences of visualizing, in contrast to the relatively clear and stable character of experiences of seeing. Now if our certainty in the truth of B survives reflecting on these points without independent reinforcement (e.g. by knowledge of a proof), that is evidence that the belief is not produced by observing the visualizing experience.

To summarize: one may acquire belief B by visualizing in the manner suggested and, in addition to circumstances (a), (b), and (c) of the previous section, it can happen that (d) the phenomenology of looking and noticing is absent, and (e) one has a feeling of certainty in B which is not undermined by recognizing the fallibility of inner observation; just as the conjunction of (a), (b), and (c) makes it extremely unlikely that, in getting belief B by visualizing in the manner suggested, one is inferring from sense experience, so the addition of (d) and (e) makes it extremely unlikely that one is performing an experiment in one's visual imagination and inferring from the visualizing experience. This is ground for thinking that it is possible to acquire belief B by visualizing in the way described without drawing on inner or outer experience as evidence.

These points do not imply or even suggest that we *never* arrive at geometrical beliefs by experiments of visual imagination. On the contrary, we sometimes do, I think. For example, one can try to discover how many edges an octahedron has by imagining a wire model of an octahedron, the straight parts of wire representing its edges, and "counting the straight parts" to get an answer. Or presented with two objects with complex shapes we can visualize rotating one to see whether it is congruent to the other.[9] Nor am I saying that in getting belief B by visualizing as suggested, one could not be performing an inner experiment and accepting the truth of B as a result (although I think it unlikely).

However, if in getting B by visualizing one *were* performing an inner experiment of visual imagination and using the visualizing experience as evidence for B, this could not be a case of genuinely discovering the truth of B. This is because, if we observe the visualizing experience, we would observe, not the perfect geometrical forms represented by phenomenal features of the visualizing experience, but the representing features themselves. We have no reason to believe that such features instantiate perfect geometrical forms, and in particular no reason to think that the phenomenal representations of squares in visual imagination are themselves perfect squares. And we clearly could not observe such perfection, since there are divergences from perfection too fine to be observed. Hence we literally cannot observe even one instance of B to be true by observing the subjective phenomena of our own visual imagery.

It follows that if one can discover the truth of B (rather than merely come to believe it) by visualizing in the manner suggested, it must be possible that this process does not constitute a way of drawing on inner or outer experience as evidence; it must be possible that it functions in some other way. Moreover, reason for thinking that there is some other way is provided by the fact that sometimes belief in B is arrived at by visualizing in the manner suggested when conditions (a) to (e) hold simultaneously.

The non-evidential role of visualizing
When one gets belief B by visualizing the corner triangles folding over onto the inner square without gap or overlap, one gets the belief B almost

immediately, that is, without any subjectively noticeable period between visualizing and getting the belief. Immediacy suggests that to explain why visualizing leads to the belief we should look to the visualizer's prior cognitive state. One hypothesis is that the subject's prior cognitive state included tacitly believing B. This kind of view was proposed by Plato. On Plato's view the experience of visualizing triggers retrieval of the tacit belief B.[10] The problem facing Plato's view is this. A necessary condition of tacitly believing a proposition is that the believer would assent to the proposition when explicitly considering it. But people have come to believe B by visualizing in the manner described, who were unable to say whether B is correct prior to visualizing, though they understood the question perfectly well. Such people, therefore, did not tacitly believe B, and so in their case the role of visualizing could not be to trigger retrieval. We want an account that covers this case too.

However, there are alternatives to Plato's hypothesis. Even if the subject's prior cognitive state did not include believing B already, it might have included resources sufficient to produce belief B upon visualizing. We can draw on the cognitive resources mentioned in the previous two chapters. Here I am assuming that visualizing and visual perceiving share many mechanisms. In fact there is substantial evidence from different sources for this. One kind of evidence concerns the effects of visualizing on performance in visual perception tasks. In particular, visualizing has been found to hinder visual perception when the visual image and the stimulus to be detected are dissimilar, and to facilitate visual perception when they are similar.[11] Another kind of evidence comes from patients with brain damage, showing deficits in imagery abilities that coincide with deficits in perception. For example, patients who fail to notice visible objects of one side of their visual field, a condition known as unilateral visual neglect, may also be unable to visualize objects on the same side of a familiar scene.[12] Yet another kind of evidence comes from brain scan studies, indicating that substantial areas of the brain involved in visualizing are involved in visual perceiving.[13] Based on data of these kinds an integrated theory of visual perception and visual imagining has been developed by Stephen Kosslyn.[14] A central and well-substantiated claim of the theory is that the stored visual category representations that are activated in perceptually recognizing the category of objects seen in unfavourable conditions, such as poor light or partial occlusion, are also activated in generating visual images of objects of a given category. Thus the stored category representation for visual recognition of squares, which might consist in something like the category specification for squares given in Chapter 2, is activated not only in recognizing a seen figure as a square but also in generating a visual image of a square in the absence of any seen figure.

Now returning to the resources that might be used in producing belief B, we can take it that relevant items of the subject's prior cognitive state include the visual category specifications that one must access to recognize a configuration of lines as a square or as a triangle, and corresponding geometrical concepts. Also included are links in associative memory between those category specifications, verbal category labels, and geometrical concepts for squares and triangles. In addition, there are the belief-forming dispositions entailed by possessing those concepts and restricted universal quantification, as explained in the previous chapter. Finally, there are basic beliefs arrived at by the activation of such dispositions, such as the belief that the parts of a square either side of a diagonal are congruent.

One possibility using such resources is as follows. First one visualizes the configuration of squares verbally described in posing the problem, as illustrated in Figure 4.1. Then one visualizes the corner triangles fold over, and this causes visualization of a new configuration within the remembered frame of the old, as illustrated in Figure 4.2. This new configuration contains the lines joining midpoints of opposite sides of the original square. Some people are not caused to visualize this by visualizing the corner triangles fold over; but this is usually remedied by asking them to visualize the lines joining midpoints of the opposite sides. Either way, then, one comes to visualize the new figure.

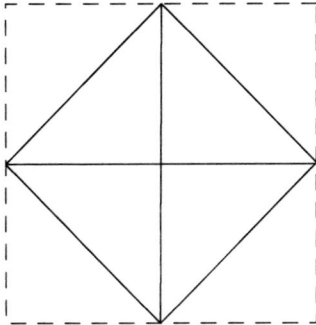

Figure 4.2

Call the lines joining midpoints of opposite sides of the original square the "cross lines" for short. The inner square is now visualized as divided by the cross lines into triangles, the "inner triangles", and the original square is now visualized as divided by the cross lines into squares, the "quadrants". The cross lines coincide with vertical and horizontal axes of the original square, and because reflection symmetry about these axes is a feature coded in the category specification for squares, we will believe that the quadrants are congruent.[15] Moreover, we will believe that the quadrants are squares. The belief that the quadrants are congruent squares may result from the activation of prior belief-forming dispositions by the visualizing. It would take too long to substantiate this, but the kind of thing I have in mind can be gathered from the previous chapter. I will take it that the belief-forming dispositions involved are reliable and are activated in circumstances that do not make it epistemically irrational to hold the belief, so that the resulting belief state is a state of knowledge.

What I have been calling the corner triangles I will now call the outer triangles. The inner and outer triangles are visualized as paired off according to the quadrants they form. Each outer triangle is visualized as the part of a quadrant to one side of a diagonal; its partner inner triangle is visualized as the part of the quadrant to the other side of the diagonal. Given that our prior cognitive state includes the beliefs that the quadrants of a square are squares and that the parts of a square either side of a diagonal are congruent, we will be prompted by the visualization to think that each outer triangle can be arranged (by a rigid transformation) to fit exactly onto the inner triangle in the quadrant to which it belongs. But the inner triangles are visualized as composing the inner square without gap or overlap; hence we will be led to think that the outer triangles can be arranged to fit onto the inner square, without gap or overlap and without any change of size or shape.

Although the visualizing has the subjective character of focusing on a particular pair of squares (one with its vertices coinciding with midpoints of the sides of the other), the belief acquired is about *any* pair of squares so arranged, and the visualizer will not think otherwise, unless she regards some feature of the visual image or its transformation as restricting its representational scope—but there is no cause for that here. This is possible because one has dispositions to acquire general beliefs that can be activated by particular visual experiences, as explained in the previous chapter. There is a salient difference between the belief-forming dispositions at work here and those discussed in the previous chapter: the latter issue directly from possession of certain concepts, whereas the former include some that we have as a result of prior beliefs. But that concerns the origin of the dispositions, not the character of the beliefs that result from their activation. Thus one can acquire the general belief that for any square c, if c_i is the square whose vertices are midpoints of c's sides, the parts of c beyond c_i can be arranged to fit exactly into c_i without gap or overlap, and without change of size or shape. This is B.

Examination of this proposal

The account given here of the role of visualizing in delivering belief stands in opposition to the accounts of previous sections, according to which the visualizing must be a way of delivering inner or outer experience as evidence for B, from which one then infers B. Those accounts, it was pointed out, have some weaknesses. In particular, they do not square with the possibility that circumstances (a)–(e) occur together. Does the account proposed in this section fare any better? Let us review (a)–(e).

(a) One feels that it is not the case that there might turn out to be a counter-example, and this feeling is not weakened by recognizing the fallibility of inductive generalization.

(b) The putative evidence of sense experience is meagre, but conviction is strong.

(c) The belief in B is not undermined by recognizing that the putative evidence of perceptual experience is of a kind that could not warrant that belief.

(d) The phenomenology of scrutinizing one's experience and noticing some feature of it is absent.

(e) One has a feeling of certainty in B that is not undermined by recognizing the great fallibility of inner observation.

Does the account proposed in this section square with the conjunction of (a)–(e)? On this account the belief is not based on perceptual evidence and is not reached by scrutinizing one's inner experience; hence the account is consistent with both (c) and (d).

What (a), (b), and (e) bring to our attention is the strength of conviction in B, the feeling of certainty, the feeling that a counter-example is not a genuine epistemological possibility. So far from being inconsistent with this sense of certainty, the story proposed above actually paves the way for an explanation of it. The visualizing on this account serves to draw attention to some prior beliefs and to activate some valid inferential dispositions. Provided that the visualizer feels certain about those beliefs and kinds of inference, it is likely that the feeling of certainty would be carried along to the inferred belief.

So the account proposed in this section does square with the conjunction of (a)–(e). As (a)–(e) do hold simultaneously when belief B is acquired by visualizing in the manner suggested, the role of visualizing here may sometimes be as suggested: to bring to mind prior beliefs and to activate prior inferential dispositions. I can see no other role compatible with (a)–(e) for the visualizing described earlier.

Can it be a case of discovery?

Not every way of coming to believe a truth is a way of discovering that truth: one must come to believe it independently, in a reliable way, without any violation of epistemic rationality. I assume that it is possible to arrive at belief B by visualizing in the manner just described without help from elsewhere, so that the independence condition is fulfilled. With regard to the rationality constraint, the considerations are just those presented in the previous chapter.[16] Circumstances that would make believing B irrational are in fact quite hard to come by; hence we can restrict our attention to cases in which the circumstances are not thus unfavourable. So the central question is one of reliability. If one were to arrive at belief B by visualizing, and if the process of visualizing had the non-evidential role outlined in the last section, would that be a reliable way of arriving at B? There are two approaches to this question. One is to examine the prior beliefs and belief-forming dispositions involved in this way of arriving at the belief. The other approach is to look at the ways in which visualizing can lead to error, and to check that the process involves none of those ways. This second approach will be followed here, partly in order to contrast the visual thinking described above with the kinds of fallacy responsible for the general notoriety of visual thinking in geometry.

First, there are mistakes that arise from a mismatch between the intention and the content of visualizing. Let me introduce an example. Suppose you were looking at a cube, not face on, but with one of its corners pointing straight at you. How many of its corners would be in view? To answer this, people naturally try to visualize a cube as it would appear corner on; but often what is actually visualized is a three- or four-sided pyramid from above and wrong answers are given.[17] In this case people simply fail to visualize what they intend to visualize. This type of error is important, but it does not destroy the possibility of discovery by visualizing in the manner described here. One could intend to visualize the corner triangles of a square folding over onto the inner square and fail while thinking that one is succeeding. But our question is whether, if one *does* succeed, that could be part of a reliable way of coming to believe proposition B. This question is pertinent because there is no difficulty in this particular task of visual imagination, whereas the cube task is often found to be difficult.[18]

A second type of error: one succeeds in visualizing as intended, but has a false belief about what one has visualized. For example, suppose one tries to visualize a scalene triangle with a horizontal base and a vertical side, and then a line from its upper vertex to the midpoint of its base. One may succeed but falsely believe that one is imagining a triangle with the altitude to its base, simply because one does not have a proper grasp of altitude.[19] This type of error is not relevant here: in the case at hand there is no false belief about what one visualizes.

A third type of error occurs when visualizing activates a fallacious inferential disposition. Such a disposition is not necessarily a disposition to infer something false. Suppose one visualizes a single figure of a certain sort, and this causes one to believe a proposition about *all* figures of that sort. An error of the third type occurs just when what is visualized illustrates the proposition only in virtue of some attribute not shared by all figures of the relevant sort. Here is an example. Suppose one visualizes a triangle with a horizontal base and three acute angles; then adds the perpendicular line falling from the upper vertex to the base; then embeds this figure in a rectangle on the same base with the same height, as illustrated in Figure 4.3.

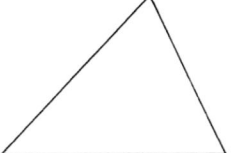 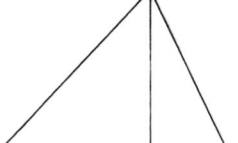 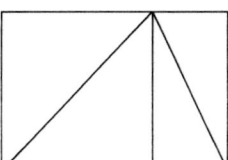

Figure 4.3

Now suppose that the visualizer comes to view the two right-angled triangles composing the original triangle as exact halves of the two rectangles that compose the encompassing rectangle. Coming to view the figure this way may cause one to believe that the area of a triangle, any triangle, equals half its altitude times its base. That would be a fallacious generalization. Though the conclusion is correct, that particular way of reaching it depends on a feature of the visualized triangle not shared by all triangles, namely, that its base angles are no greater than right angles; this is because a triangle with a base angle greater than a right angle cannot be visualized as part of a rectangle having the very same base.[20]

I suspect that this is a kind of error that is easy to make, perhaps one of the commonest types of error in visual thinking. Are you making an error of this third type if you visualize the corner triangles of a square folding over onto its tilted inner square, and as a result come to believe that the corner triangles of *any* square can be arranged to fit the tilted inner square exactly? No: if the corner triangles of one square can be arranged to fit, the corner triangles of any square can be so arranged, as all squares have the same shape, while any property independent of shape is irrelevant.

Finally images may be too faint and inconstant to produce enough confidence for a belief as opposed to a mere inclination to believe. The visualizing here has sufficiently few stages and the imagery is sufficiently simple that these problems need not arise. It is easy to rehearse the visualization task and thereby gain the support of memory in producing stronger imagery. So we can simply restrict our attention to cases in which the imagery is clear and stable.

Several types of error that beset belief-acquisition by visualizing have been canvassed, and I am not aware of any others. No error of these types is involved when we get belief B by visualizing in the manner suggested. If the visualizing were a way of drawing on experience as evidence, then (as pointed out before) certain errors of inference from experience would be involved. But if the visualizing has the non-evidential role sketched above, that route to belief B involves none of the kinds of error that visual thinking is heir to. Of course this is not absolutely conclusive, since there may be relevant types of error I have overlooked. Still, this investigation makes a favourable verdict very plausible.

GEOMETRICAL DISCOVERY BY VISUALIZING   67

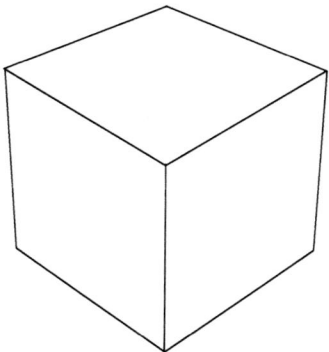

Figure 4.4

Concluding remarks

It is natural to start off with the suspicion that discovery of a general geometrical theorem by visualizing is impossible. Visualizing is most readily compared with seeing; seeing, in its primary role of providing observational evidence, cannot deliver this kind of geometrical knowledge; and it is felt that visualizing is no better than seeing.[21] But if the account given here is correct, the comparison is misleading. While the *experience* of visualizing is similar to the experience of seeing, the *epistemic role* of visualizing can be utterly different from the primary, evidence-providing role of seeing. Even seeing may on some occasions have a non-evidential role in delivering knowledge, as illustrated in the previous chapter. So the fundamental mistake here is to assume that the epistemic role of visual experience, whether of sight or imagination, must be to provide evidence. In view of its non-evidential role we can say that visualizing in this case is part of an *a priori* means of acquiring belief. If we accept that it results in discovery, we can say that it is part of an *a priori* means of discovery. Moreover, the visual element is a non-superfluous part of the process. If we delete a merely illustrative diagram accompanying a verbal presentation of a proof, going through the remaining verbal text may still produce belief. But if we delete the visualizing from the process of discovery described above, what remains would not yield belief, since the relevant inferential dispositions would not be activated. Thus the process is essentially visual.

There is no analysis of meanings, and no deduction from definitions in the process. In philosophers' jargon the process is *a priori* but not analytic;

68 GEOMETRICAL DISCOVERY BY VISUALIZING

rather it consists in the operation of a synthesis of visually triggered belief-forming dispositions. Hence it may be appropriately regarded as a synthetic *a priori* route to knowledge. Moreover, the earlier beliefs on which the process draws can be known in a synthetic *a priori* way. In the previous chapter the case was made for this claim with respect to one of these beliefs, the belief that the parts of a square either side of a diagonal are congruent. A similar case can be made with respect to the other belief, the belief that the quadrants of a square are congruent squares. Hence the target belief is knowable in a synthetic *a priori* way. So we have some reason to accept Kant's view (or a view close to it) [22] that some geometrical knowledge may be synthetic *a priori*.

Notes

1. This is not to say that proof is unimportant for mathematics. On the contrary, it is vital. The point is only that there can be discovery without proof.
2. I came across this example in Kosslyn (1983) who got it from Jill Larkin. It is similar to an example in Plato's dialogue *Meno* (Plato 1985).
3. By this route or one very close to it, one could arrive at the belief that this holds for squares in physical space. But that is not the belief under consideration here. We are considering the case in which one arrives at a belief about the kind of squares that would have been instantiated had physical space been Euclidean. I omit the qualification "Euclidean" henceforth.
4. There is evidence that the process of visualizing paper-folding is similar to that involved in seeing paper-folding when performed by oneself. Shepard and Feng (1972).
5. By induction here I mean generalization from a sample of observed instances, not mathematical induction.
6. I found to my surprise that there is such a thing as red snow, snow coloured red by the presence of certain algae and diatoms. But of course it is not really the snow that is red.
7. This kind of epistemic certainty is a feeling about one's evidential situation with respect to B; one may also feel that B has the metaphysical status of a necessary truth. Epistemic certainty and metaphysical necessity are not the same, and not even co-extensive.

8. One might respond that in arriving at B by visualizing one has to be *unconsciously* drawing on past experience as evidence for B, so that the process of belief acquisition would be protected from a conscious epistemic conviction that the experience would be evidentially insufficient. Maybe; but were we to reflect on the resulting belief and our reasons for it, it would lose protection, and so be undermined.
9. Shepard and Metzler (1971).
10. *Meno* 81e–86c; *Phaedo* 72e–73b. See Plato (1985) and Plato (1993) respectively.
11. This is a slight oversimplification. For a review of the behavioural evidence see Finke (1989) and Kosslyn (1994: ch. 3).
12. Bisiach and Luzzatti (1978). For a review of this kind of evidence see Farah (1988).
13. Kosslyn et al. (1997) found that about two-thirds of the brain areas activated during visual imagining and visual perceiving are activated in common. For a review of this kind of evidence see Thompson and Kosslyn (2000).
14. Kosslyn (1994).
15. The category specification for squares is given on p. 23. Reflection symmetry entails that each quadrant is congruent to each of its adjacent quadrants; from this and the transitivity of congruence it is clear that opposite quadrants are congruent. I assume that one already believes that congruence is transitive.
16. See the subsection headed "Rationality" in the section headed "Is it knowledge?"
17. The wrong answers often given are four and five. The correct answer is seven. See Figure 4.4. The example is taken from Hinton (1979).
18. I have been asked whether and how we *know* that we succeed in visualizing the corner triangles fold over onto the inner square. But this is beyond my present concern; for it is not the case that to have made a discovery one must know that one has done so.
19. See the constructions of student B in Hershkowitz (1987).
20. Of course, one can make a special provision for these cases: imagine the triangle in a parallelogram whose diagonal is the triangle's longest side and whose base is the triangle's base. Then the area of the triangle will be half that of the parallelogram. The conclusion can then be inferred

from the fact that the area of the parallelogram is its base times its perpendicular height. Cf. Chapter 8, below.
21. The influential philosopher Philip Kitcher writes in this vein, when arguing against the idea that "intuition" could be a means of mathematical knowledge. Kitcher (1984: ch. 3, §§ I, II).
22. Some caution is needed in making comparisons with Kant's views. See ch. 3 n. 15.

5
Diagrams in Geometric Proofs

The concern of the previous chapter was the possibility of discovering a geometrical truth by means of visual imagination. This chapter turns from discovery to justification. Some ways of discovering a truth provide not only knowledge but also a justification for belief, as when one discovers something by constructing a proof of it. Often, however, discovery is not coincident with justification, as in the kind of discovery illustrated in the previous chapter. In that case one may want the extra assurance provided by a justification, especially if one is aware of having reached the new belief by means that are easy to misuse. It is well known that overgeneralization is a hazard of visual thinking without due care, as one's reasoning may depend inconspicuously on some feature represented in the image or diagram that is not common to all members of the class one has in mind. So if we have used visual means in making what we think is a geometrical discovery, we may very well seek to justify or check our conviction by proving the proposition concerned.

Presentations of geometric proofs are often accompanied by diagrams for quick and easy comprehension. But to many people it seems clear that diagrammatic reasoning cannot be a part of the argument itself, otherwise it would be prey to the very insecurity that we want to eliminate, insecurity from visual thinking, and so the argument would not be able to justify its conclusion; it would not be a proof. This is the line of thought that I think most strongly supports the widespread belief that diagrams can have no epistemological role in proof. The main aim of this chapter is to investigate this negative view and the argument for it just presented.

Before proceeding, some distinctions should be laid out to reduce confusion. First, we must distinguish between a proof and a presentation of a proof. Presumably one and the same proof can be presented in different languages, or with different wording in the same language. But we have to face up to an essential vagueness here. How different can distinct presentations be and yet be presentations of the same proof? There is no context-invariant answer to this, and even within a context there may be some indeterminacy. Usually mathematicians are happy to regard two presentations as presenting the same proof if the central idea of the proof is the same in both cases. Consider an example from number theory. In some modern textbooks what is dubbed Euclid's proof of the infinity of primes is presented as a proof by contradiction from the supposition that there is a largest prime. But the argument as presented in Euclid's *Elements* (Book IX, Proposition 20) is not a proof by contradiction.[1] It does contain as a subpart an argument by contradiction, not from the supposition that there is a largest prime, but from the supposition that the successor of the product of a set of primes has a prime factor that is one of the primes in the set. That little sub-argument, together with appeal to the theorem that every integer greater than 1 has a prime factor, is common to both presentations. As this common part contains the pivotal idea, mathematicians think of the ancient and modern texts as presenting the same proof. This, I guess, is because mathematicians are concerned with providing proofs, and for that end finding a pivotal idea is the hardest and most interesting aspect.

But if one's main concern is with what is involved in *following* a proof, as relevant as the proof's central idea are its overall structure and its sequence of steps. In that context the ancient and modern texts present distinct proofs. The concern in this chapter is with cognitive processes involved in following a proof, where that includes both reading a presentation of an argument appreciating its cogency at each step, and thinking through an argument appreciating its cogency without a textual presentation. Hence what is needed here is a fairly fine-grained individuation of proofs. Even so, not every cognitive difference in processes of following a proof will entail distinctness of proofs: presumably the same information in the same order can be presented in ink and in Braille so the same proof may be followed using different cognitive abilities.

Once individuation of proofs has been settled, another pertinent distinction can be introduced. In the process of following a proof, a given part of the thinking is *replaceable* if there is thinking of some other kind whose substitution for the given part would result in a process of following the same proof. To make the idea of replaceability a little more concrete, suppose in following a proof one extracts information from a diagram that accompanies the verbal text of a presentation of the proof. Now suppose the diagram is cut out and the verbal text is extended to convey just the information that had been extracted from the diagram in following the argument. Following the argument in this modified form may match following the argument of the original presentation in all respects relevant for proof individuation; in that case it is the same proof that is followed in the two processes. So that part of the original process that constitutes extracting information from the diagram is replaceable. Derivatively, we may say that the diagram itself is replaceable in the original presentation of the proof.

Replaceability should be distinguished from superfluity. In the process of following a proof, a given part of the thinking is *superfluous* if its excision without replacement would result in a process of following the same proof. Suppose a verbal text and an accompanying diagram presents a proof, but any information relevant to following the proof that can be extracted from the diagram is already conveyed in the verbal text. Suppose also that prior to reading the text one looks at the diagram, and that at certain stages in following the argument as presented in the text one looks back at the diagram to check one's understanding of the sentences characterizing the situation to be considered. Those parts of the total process that involve looking at the diagram facilitate understanding the verbal text and help to confirm that understanding. But if, as is possible, one's reading of the text alone provides all the relevant information for following the proof, those parts of the process that involve looking at the diagram are superfluous. Derivatively, we may say that the diagram itself is a superfluous part of the presentation.

Few will deny that there can be superfluous diagrammatic thinking in following a proof. Like Tenniel's illustrations of scenes in Lewis Carroll's *Through the Looking Glass*, mathematical diagrams can be stimulating without providing extra information relevant to the narrative: one can get the whole story by reading the text. This leaves a number of possibilities.

(1) All thinking that involves a diagram in following a proof is superfluous.
(2) Not all thinking that involves a diagram in following a proof is superfluous; but if not superfluous it will be replaceable.
(3) Some thinking that involves a diagram in following a proof is neither superfluous nor replaceable.

74 DIAGRAMS IN GEOMETRIC PROOFS

The negative view stated earlier that diagrams can have no role in proof can be more sharply recast as (1). I shall try now to present a counter-example.

Non-superfluous diagrammatic thinking?

To test the claim that diagrammatic thinking is always superfluous, I will consider a particular example. Here is a straightforward proof of a simple theorem of Euclidean plane geometry:

> For any circle and any diameter of it, the angle subtended by chords from opposite ends of the diameter to a common point on the circumference is a right angle.
>
> Let BD be a diameter of any circle with centre C. Let AB, AD be the two chords from the endpoints of diameter BD to any point A on the circumference. Let γ and δ be the sizes of ∠ BCA and ∠ DCA respectively. These are the data. (See Figure 5.1.)

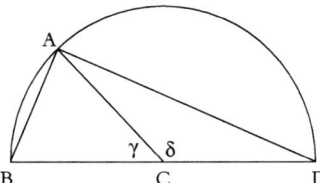

Figure 5.1

1. The angle on one side of a diameter of a circle about its centre is a straight angle. [premiss]
2. The angle on BD about C on the same side as A is a straight angle. [1, data]
3. The angles ∠ BCA and ∠ DCA compose the angle on BD about C on the same side as A. [data]
4. The angles ∠ BCA and ∠ DCA compose a straight angle. [2, 3]
5. The angles composing a straight angle sum to 180°. [premiss]
6. γ + δ = 180. [4, 5, data]
7. Two radii of a circle (not forming a diameter) are equal sides of an isosceles triangle. [premiss]
8. CB and CA are radii of the circle (not forming a diameter). [data]
9. The angles opposite equal sides of an isosceles triangle are equal. [premiss]

10. ∠ BAC = ∠ ABC. [7, 8, 9, data]
11. The internal angles of a triangle sum to 180°. [premiss]
12. ∠ BAC + ∠ ABC + γ = 180. [11, data]
13. ∠ BAC + ∠ ABC = 180 − γ = δ. [6, 12]
14. ∠ BAC = δ/2. [10, 13]

Noting that CA and CD are radii (not forming a diameter), we find by parallel reasoning that

15. ∠ DAC = γ/2.
16. ∠ BAC + ∠ DAC = δ/2 + γ/2 = (δ + γ)/2 = 90. [6, 14, 15]
17. Angles ∠ BAC and ∠ DAC compose angle ∠ DAB. [data]
18. ∠ DAB = 90. [16, 17]
19. An angle of 90° is a right angle. [premiss]
20. ∠ DAB is subtended by the chords AB and AD from A on the circumference of a circle to the ends B, D of a diameter. [data]
21. The angle subtended by chords from a common point on the circumference of a circle to opposite ends of a single diameter is a right angle. [18, 19, 20]

This presentation is fairly explicit, much more explicit than normal. But it is not totally explicit. To reach (6), the claim that γ + δ = 180, it is first stated that angles ∠ BCA and ∠ DCA compose the angle on BD about C on the same side as A, which is (3). How do we know that this (3) is true? Although the text indicates that it is given, it is not among the situation data explicitly stated. But somehow it is clear to us. One way in which this might happen is as follows. We see the diagram in a certain way, namely, as a diagram of a semicircle with two chords from a point A on the circumference to the endpoints B, D of the diameter. Seeing this, and discerning from the positions of the letters which elements of the depicted structure are named by which letters or letter sequences, we inspect the smallest part of the diagram containing representations of the elements mentioned in the claim, the part reproduced in solid lines in Figure 5.2, to see that the claim is correct.

Here I am assuming that one is using certain conventions for extracting information from the diagram part, conventions that would have been used in constructing the diagram. Some of this information is as follows: the angles ∠ BCA and ∠ DCA are non-overlapping angles on one side of BD about C; they are on the same side of BD as A; together they compose the

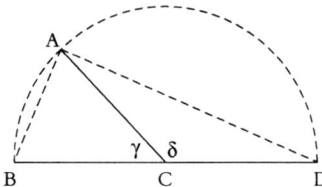

Figure 5.2

entire angle on that side of BD about C. This gives us (3). Can one tell that (3) is correct from an understanding of the text which excludes any use of the diagram? Yes, but only if one uses some other bridge from the explicitly stated data to the claim. We could instead apply rules governing the use of letters and letter sequences naming elements of the structure specified by the data. Here, for example, we could apply:

> If XZ is a segment with interior point Y and W is a point not on the line containing XZ, angles XYW and ZYW compose the angle on XZ about Y on the same side as W.

The important facts now are these. First, some way of reaching the information of (3) from the explicit data is needed in order to follow the proof, as that information is not among the explicit data. Secondly, there are at least two ways of bridging the gap between the explicit data and the information of claim (3): tacit visual extraction of information from a diagram, and tacit application of rules about letter sequences. In following the proof as presented, one gets to (3) from the data by extracting information from the diagram. That part of the process, extracting information from the diagram, is not superfluous, as some bridging is needed. But it is replaceable, as the bridging could be achieved by applying schematic rules about letters and letter sequences.

Parallel remarks apply to one of the unstated assumptions used in reaching (10) from earlier claims and the data. This is the assumption that angles ∠ BAC and ∠ ABC are angles opposite the sides CB and CA of triangle ABC. We could reach this by applying some rule on letter sequences, such as:

> In a triangle XYZ, the angle opposite the side denoted by the concatenation of two distinct letters PQ from {X,Y,Z} is denoted by the angle sign ∠ followed by the concatenation of any three distinct letters from {X,Y,Z} whose second letter is neither P nor Q.

Alternatively we can extract this information by inspecting part of the diagram containing representations of the elements mentioned (as in Figure 5.3), taking the angle *opposite* a given side of a triangle to be the internal angle whose arms stretch from a common vertex to the endpoints of the given side.

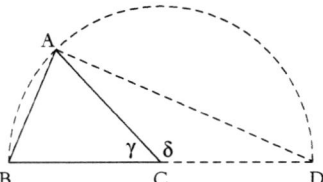

Figure 5.3

If we do get the information that ∠ BAC and ∠ ABC are angles opposite the sides CB and CA by inspecting the diagram, that bit of visual thinking will be replaceable but not superfluous.

If, as seems plausible, the geometrical argument really is a proof, these considerations refute the view that all diagrammatic thinking in the process of following a proof is superfluous. What prompts this view is the belief that diagrammatic thinking is epistemically insecure, and that if some diagrammatic thinking is a non-superfluous element of a cognitive process, the insecurity of the diagrammatic element entails that the process cannot be one of following a proof. The fallacy of this argument will be easier to spot if we recall the kind of error that diagrammatic thinking is prone to. The danger of diagrams is that they tempt one to make unwarranted generalizations, as one's thinking may too easily depend in an unnoticed way on a feature represented in the diagram that is not common to all members of the class one is thinking about. Recall an example from the previous chapter: using a diagrammatic construction that is only possible for triangles whose base angles are both acute, the conclusion drawn is that the area of *any* triangle equals its height times its base—a clear case of unwarranted generalizing. But in the cases examined above, no generalizing occurs. Consider, for example, the first case, in which one extracts from the diagram the information that angles ∠ BCA and ∠ DCA compose the angle on BD about C on the same side as A. That is a claim about the arbitrary instance introduced by the initial "Let" sentences that supply the data, not a claim about all figures of a certain sort. In these cases of diagrammatic thinking, rather than *overlooking* features represented in the diagram on which the reasoning depends, one *notices* relevant features and then records them explicitly. Given that there is no danger of misreading the diagram, there is no insecurity in a step in thought of this kind. So the argument for the view that diagrammatic thinking must be superfluous has a mistaken premiss, amusingly itself an overgeneralization, namely, that all diagrammatic thinking is epistemically insecure.

Valid generalizing with diagrams

All that is fine, but it does not settle the matter. For it leaves intact a slightly refined version of the negative view of diagrams: the charge is not that all non-superfluous use of diagrams is epistemically insecure; rather, it is that generalization is insecure when the reasoning that leads up to it involves the non-superfluous use of a diagram.

The argument given earlier contains a step of generalization, at the end. Perhaps in making this step one is unwittingly using or depending on the diagram in a fallacious way. To look into this we first need to be clear how any generalizing can occur in a proof. Of course generalizations abound in science: the step from observational data to a general claim that explains those data may be epistemically acceptable, absent an alternative explanation as good or better. But that kind of generalizing is always a step from premisses to a conclusion that is not deductively entailed by those premisses, hence cannot occur in a mathematical proof. Generalizing occurs in mathematical argument when one reasons about an arbitrary instance of a class of things in order to draw a conclusion about all members of the class. Here is a schematic illustration. On the right of each line I indicate the earlier lines from which it is obtained if it is not a premiss; on the left I indicate the premisses on which that line depends.

{1}	1. Let c be any K.	[premiss]
{2}	2. Every K has E.	[premiss]
{3}	3. Every K that has E has F.	[premiss]
{1, 2}	4. c has E.	[1, 2]
{1, 3}	5. If c has E it has F.	[1, 3]
{1, 2, 3}	6. c has F.	[4, 5]
{2, 3}	7. If c is a K, c has F.	[6]
{2, 3}	8. Every K has F.	[7]

Typically, some general claims about the class, call it K, will be used as premisses, and there will be a "Let" sentence introducing an arbitrary instance of K. (If there are several classes involved, as is usual, there will be several such "Let" sentences. For simplicity we will ignore this.) Then using the general premisses some conclusion is drawn about the arbitrary instance, say that it has property F, via some intermediate steps. In the example the general premisses are lines 2 and 3, and the conclusion about an arbitrary K is line 6. In the next line this conclusion is weakened by making it conditional on the content of the initial line, as in line 7.[2] Hence this conditional does not depend on the initial line as a premiss: the premiss, we say, is *discharged*. Then the general conclusion is drawn that every K has the property, as in the step from 7 to 8. This is valid provided that the property F can be specified without reference to c and that there is no mention of c in any of the premisses that the line from which the generalization is obtained depends on. In this case these are the undischarged premisses 2 and 3 used in reaching line 7. This sort of inference is known in logic textbooks as universal generalization, or UG for short. Appendix 5.2 presents a formal version of UG and a proof of its validity.

The question we need to answer is whether this kind of valid generalizing can occur in an argument that uses a diagram in a non-superfluous way to reach an intermediate statement about an arbitrary instance. The worry is that in using a diagram to reason about an arbitrary instance c of class K, we will be using some feature of c represented in the diagram that is not common to all instances of the class K. In the example of the previous chapter, we used the fact that the triangle in the diagram was represented as contained in the rectangle having the same base line and the same height as the triangle. This is an unstated premiss mentioning the arbitrary instance[3] (the triangle represented in the diagram), which thereby violates one of the essential conditions for universal generalization.

Is there a parallel error in our example, the diagrammatic argument that the angle subtended by a diameter of a circle to any point on its circumference is a right angle? The diagram represents part of an arbitrary instance of the class of circles with chords from opposite endpoints of a diameter meeting at a point on the circumference. What property of the instance not shared by all members of the class might we be unwittingly relying on when following the argument? Well, there is the size of the semicircle containing the chords' meeting point and there is the orientation of the semicircle. But it is clear that the argument does not depend in any way on those features. There is also the position of the chords' meeting point relative to the diameter's endpoints, and with that there are the following properties: the ratio of chord lengths BA to AD; the ratio of arc lengths BA to AD; the ratio of angles γ to δ. A simple step by step inspection makes it clear that none of these properties is relied on in the argument. In particular, there is no dependency on the represented inequality of γ and δ. As there is no threat from any other properties that the figure is represented as having, the argument contains no violation of the conditions for deductively valid generalizing. I conclude that valid generalizing can occur in an argument that uses a diagram in a non-superfluous way to reach an intermediate statement about an arbitrary instance.

A different objection: the transparency of proof

One response is to concede that point, but object as follows. A diagram may be used in a non-superfluous way in a sound generalizing argument; but the argument cannot be a proof. When a diagram is used non-superfluously in such an argument, we need in addition to the argument itself an inspection of its steps, as shown in the preceding discussion, to ensure that it contains no illegitimate dependency on a special feature represented in the diagram. Without an inspection of this kind, we do not have the degree of rational assurance needed to justify believing the conclusion, as the argument lacks the requisite degree of transparency; and so the argument does not constitute a proof. The objection continues: this holds not just of this particular argument but of all arguments in which there is non-superfluous use of a diagram to represent an instance and generalization from a fact deduced about that instance. This is because no diagram of an instance can avoid representing it as having a property not shared by all members of the relevant class. For example, a diagram for the argument given earlier cannot avoid representing the angles γ and δ as unequal or as equal, whereas a purely verbal description of the instance can leave it unspecified whether they are equal or not. For this reason an argument containing a generalization from a diagram (as I shall call such arguments[4]), though a valid argument from premises known to be true, cannot be a proof. This is the objection.

This, I find, is a weighty objection. Its weakest point is the claim that arguments containing a generalization from a diagram do not have the degree of transparency required for proof. The degree of transparency here is the degree to which it is clear exactly what are the premises and the steps in the argument. The problem with the objection is that there is no fixed degree of transparency required for proof. It is a matter of purpose and context. In some contexts nothing short of the utmost transparency will do. That may mean that only a formalized argument will be counted a proof, and only when the system of axioms and inference rules deployed in the argument is known to be sound. At the other pole, research mathematicians can think through and communicate their proofs to one another in a quite casual way, relying on what they know to be common knowledge in the research community, often indicating their intentions with diagrams. Both their thinking through the argument and their presentation of it will have a relatively low degree of transparency. So the objection needs to be refined and elaborated in one of two ways: either it must specify a type of context (or purpose) and show that in that context (or for that purpose) arguments containing a generalization from a diagram are insufficiently transparent to count as proofs; or it must show that in no context whatever is such an argument sufficiently transparent to count as a proof.

In neither way can the objection be made good. Let us look at the bolder version of the objection, namely, that in no context is an argument containing generalization from a diagram transparent enough to be a proof. This entails that any diagram-free argument that has the same lack of transparency as an argument containing generalization from a diagram also fails to be a proof. But the kind of error that we sometimes make in generalizing from a diagram we are also prone to make when generalizing without a diagram. Often we unwittingly rely on a feature of the instance that we are using as the arbitrary exemplar of the relevant class, for we naturally tend to think of a *typical* member of the class, thereby ignoring unusual, limiting, and "pathological" cases.

To see how easy it is to make this kind of error when no diagram is in play, consider an amusing argument invented some years ago that can now be found on web pages of mathematical humour. It is an argument for the proposition that all horses are the same in colour. It is an argument by mathematical induction on the positive integers for the claim that for any positive integer n, all the members of any set of n horses are the same in colour. If this conclusion were correct, all horses would be of the same colour, as the actual population of horses constitutes a set of n horses for some positive *n*. In this case our "colour" classifications are artificial and gross, and there is no danger of indeterminacy or overlap, and we can assume that sameness in colour is a transitive relation. To reach the conclusion by mathematical induction we only need to prove that (a) it holds in the particular case when $n = 1$, and (b) for any positive integer *m*, it holds for $m + 1$ if it holds for *m*.

> *Proof of* (a). First, consider any 1-membered set S of horses. As S contains only 1 horse, it is trivially true that all the members of S are the same in colour. Thus (a) has been shown.
>
> *Proof of* (b). Assume that all the horses in any set of *m* horses are the same in colour, where *m* is any positive integer.[5] Now consider any set of $m + 1$ horses. Remove any one of the horses and we are left with a set S of *m* horses, which, by the assumption, are all of the same colour. Now swap the horse that was originally removed—call it "Lucky"—for one of the horses in the set S of *m* horses, to get another set S′ of *m* horses. Again by the assumption, all the horses in S′ are the same in colour; so in particular Lucky is the same in colour as all the rest of the horses in S′, and they are the same in colour as the horse for which Lucky was swapped, as they were together with that horse in set S. So all the horses in the original set of $m + 1$ horses are the same in colour. Hence for any positive integer *m*, if all the horses in any set of *m* horses are the same in colour, all the horses in any set of $m + 1$ horses are the same in colour. Thus (b) has been shown.

The conclusion that all the horses in any set of horses, however numerous, are the same in colour follows by mathematical induction.

With all the priming given earlier, you will probably have located the fallacy right away. But without help it often takes people a while to see where the argument goes wrong, a fact which testifies to a lack of transparency in this kind of thinking, despite the absence of a diagram. The error lies in the argument for (b). It does not work when *m* is 1. For in that case we start with a set of two horses, Lucky and one other, call it "Happy". Remove Lucky to get the set S, whose sole member is Happy. Do the swap to get the set S′, whose sole member is Lucky. Now we say that Lucky is the same as all the rest of the horses in S′, and all the rest of the horses in S′ are the same in colour as Happy, ensuring by transitivity of sameness in colour that Lucky is the same in colour as Happy. But there are no horses in S′ other than Lucky, so "the rest of the horses" does not supply the middle term needed for an application of transitivity.

Why are we prone to miss this error at first hearing? I think that the explanation lies in the representation we use of a set of size m. We ignore atypical set sizes (cardinal numbers). For every positive integer m except 1, the members of a set of size m form a plurality. The singular-plural distinction is highly marked in language and thought, and so 1 is cognitively atypical. Another cognitive fact is also likely to be at work in this case. Experiments have shown that we can rapidly and accurately detect the number in a random array of dots or a rapid sequence of sound pulses without conscious counting if that number is less than 5. For larger numbers we take longer and are more prone to error. The rapid and accurate detection of cardinal numbers from 1 to 4 is known as *subitizing*.[6] There is a subjectively salient difference between the subitizable numbers and the rest, and this is marked in language.[7] As integers from 1 to 4 seem special, we do not treat them as typical numbers. Thus our representation of a typical m-membered set, though not completely determinate, will exclude its being a set of 4 or fewer items. The same is true for our representation of a typical set of $m + 1$ horses. So for any two horses in our typical set, the Lucky and the Happy of that set, there will be others in the set besides those two.[8] It is this that enables the fallacious thinking to glide by so easily.

In this case, where no diagram and no visual thinking is used to follow the argument, there is exactly the same kind of error sometimes found and often remarked upon in generalizing from a diagram. The argument here depends in an easily overlooked way on a feature of our representation of an $(m + 1)$-membered set that is not common to all sets of more than one member, namely, that the number is beyond the subitizing limit, that is, greater than 4. So arguments containing a step of universal generalization with no use of a diagram are not in general more transparent than arguments containing generalization from a diagram. If those with generalization from a diagram must fail to be proofs because of a lack of transparency, so must those with diagram-free generalization. But this is implausibly extreme. So we should accept that some arguments containing generalization from a diagram, such as the one presented earlier in this chapter (which has no fallacy or falsehood), are at least in some contexts sufficiently transparent to count as proofs.

Now let us turn to the more modest version of the objection: in *some* context(s) any argument containing a generalization from a diagram will be insufficiently transparent to count as a proof. This entails that in contexts that require the highest degree of transparency, arguments containing generalization from a diagram do not count as proofs. The highest degree of transparency is achieved by formalized arguments, that is, derivations in an interpreted formal system. In those cases, the thinking is split into two parts, interpreting the formalism and following a formal derivation. Deductive reasoning is replaced by following a formal derivation, and that consists in the performance of a sequence of perceptual tasks: at each step we must perceive that the next symbol configuration is not merely well formed but also obtainable from earlier well-formed configurations according to specified rules of symbol manipulation. By the nature of formal systems, each step in a formal derivation is an utterly transparent symbol change: deleting, adding, reordering, separating, concatenating, or any combination of these according to a finite set of precise, determinate rules. Following a derivation delivers knowledge only against a background of (a) an interpretation of the formalism (otherwise the conclusion of the derivation is just a symbol configuration); (b) knowledge that the undischarged premises of the derivation under that interpretation are true; (c) knowledge that the derivation system is sound with respect to the formal semantics of the system. Given such a background, formal derivations provide us with the highest degree of rational assurance.

It follows that the objection in its more modest form entails that no interpreted formal system known to be sound involves the non-superfluous use of diagrams to arrive at general truths. This is false. A formal diagrammatic system of Euclidean geometry called "**FG**" has been set out and shown to be sound by Nathaniel Miller.[9] For the sake of illustration, Figure 5.4 is Miller's derivation in **FG** of Euclid's first theorem that on any given finite line segment an equilateral triangle can be constructed.

Of course, the actual argument depends on the syntax and semantics of the system, which I will not stop to present. But on a pretty obvious construal the derivation gives a fair representation of the visual steps one might take in following Euclid's own argument.[10] I doubt that it is possible to follow that argument without some visual thinking.[11] Certainly one cannot follow this argument in **FG** without visual thinking. We can conclude that there can be sound arguments for conclusions reached by

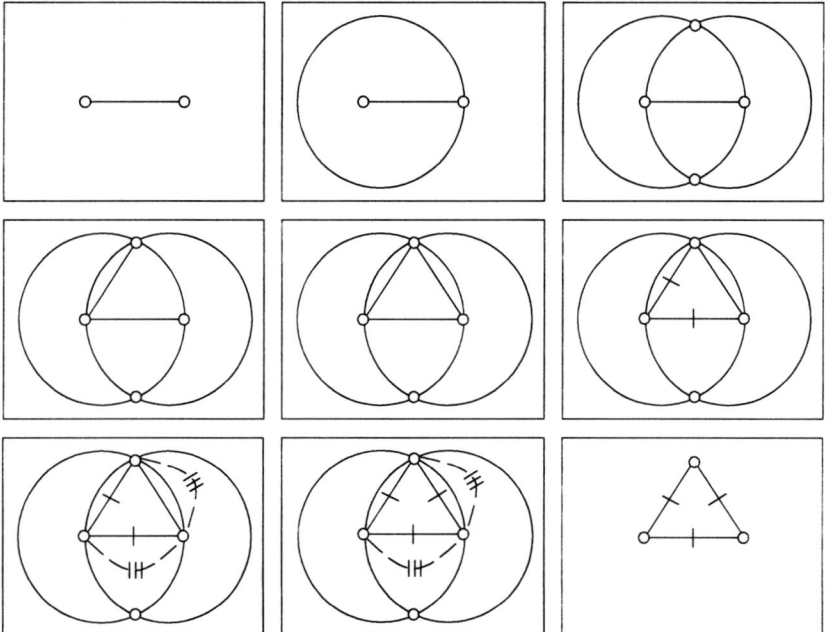

Figure 5.4

generalizing from non-superfluous diagrams with sufficient transparency to count as proofs, even when the most demanding standards are required. Moreover, diagrams are quite trivially irreplaceable (except perhaps by other diagrams) in arguments using **FG**, so some thinking that involves diagrams in following a proof is neither superfluous nor replaceable.

Conclusion

Granting that there can be superfluous use of diagrams in following a proof, there are just three exclusive possibilities:

(1) All thinking that involves a diagram in following a proof is superfluous.
(2) Not all thinking that involves a diagram in following a proof is superfluous; but if not superfluous it will be replaceable.
(3) Some thinking that involves a diagram in following a proof is neither superfluous nor replaceable.

The foregoing considerations show or at least make it very credible that the third of these is correct. But my chief concern has been to refute the first of these and its main buttress, the claim that no argument that includes generalizing from a diagram is a proof. The negative view of diagrams in geometry encapsulated in these claims is unjustified and incorrect. Still, in a given context where justifications are required there are certain standards of cogency and transparency that an argument, with or without diagrams, must meet in order to be a proof. Often, when diagrams are used, those standards are not met. The practical lesson to be drawn from our awareness of the pitfalls of diagrams in geometry is not to avoid diagram-dependent arguments altogether, but to be careful and if necessary carry out a subsequent check.

Notes

1. See App. 5.1 for a comparison of the proof (of IX. 20) as presented in Heath's translation of Heiberg's edition of Euclid's *Elements* (Euclid 1926) with the proof as presented in Burton (1980).
2. In this context the word "content" is used contrastively with "mood". The content of the sentence "Let c be any K" is just that c is a K; its mood is optative. Strictly speaking, of course, the illustration uses sentence forms rather than sentences with content.
3. Mentioning an arbitrary instance is quasi-reference, as there is no entity *e* such that one refers to *e* in mentioning an arbitrary instance. What it amounts to at the level of thought is a simulation of referring to an instance without paying attention to which instance this is. Parallel remarks apply with regard to phrases such as "reasoning about an arbitrary instance".
4. These are arguments containing (i) non-superfluous use of a diagram representing an instance, (ii) deduction of a statement about that instance, and (iii) generalization from that statement.
5. This assumption is known in the jargon as the inductive hypothesis. We use it to deduce the same for $m+1$, and then we discharge the assumption by conditionalizing, i.e. we conclude that *if* it holds for m it also holds for $m+1$.
6. There is a large literature on subitizing. Influential studies are Chi and Klahr (1975) and Mandler and Shebo (1982).
7. Hurford (2001).

DIAGRAMS IN GEOMETRIC PROOFS 87

8. So what size do we represent the typical set as having? In my view we represent it as having many members, a number greater than 4, but there is no particular number n such that we represent it as having exactly n members. I discuss this kind of representational indeterminacy in Ch. 8.
9. Miller (2001).
10. This is as intended by Miller. See Euclid (1926: Proposition 1.1).
11. Norman (2006) shows that diagrammatic thinking is needed in following Euclid's famous argument that the internal angles of any triangle sum to two right angles (Proposition 1.32), and argues cogently that this thinking contributes to the justification that is conferred on the theorem by the argument.

Appendix 5.1. Two presentations of a proof of the infinity of primes

Euclid's Presentation as translated by Heath (Euclid 1926)

Book IX, Proposition 20: *Prime numbers are more than any assigned multitude of prime numbers.*

Let A, B, C be the assigned prime numbers; I say that there are more prime numbers than A, B, C. For let the least number measured by A, B, C be taken, and let it be DE; let the unit DF be added to DE. Then EF is either prime or not.

First, let it be prime; then the prime numbers A, B, C, EF have been found which are more than A, B, C.

Next, let EF not be prime; therefore it is measured by some prime number [Book VII Proposition 31]. Let it be measured by the prime number G.

I say that G is not the same with any of the numbers A, B, C.

For, if possible, let it be so.

Now A, B, C measure DE; therefore G also will measure DE. But it also measures EF.

Therefore G, being a number, will measure the remainder, the unit DF: which is absurd.

Therefore G is not the same with any one of the numbers A, B, C.

And by hypothesis it is prime.

Therefore the prime numbers A, B, C, G have been found which are more than the assigned multitude of A, B, C. QED.

David Burton's Presentation (Burton 1980)

Theorem 3-4 (Euclid). *There are an infinite number of primes.*

Proof: Euclid's proof is by contradiction. Let $p_1 = 2, p_2 = 3, p_3 = 5, p_4 = 7, \ldots$ be the primes in ascending order, and suppose that there is a last prime; call it p_n. Now consider the positive integer $P = p_1 p_2 \ldots p_n + 1$.

Since $P > 1$, we may put Theorem 3-2 to work and conclude that P is divisible by some prime p. But $p_1, p_2, \ldots, p_n$ are the only prime numbers, so that p must be equal to one of $p_1, p_2, \ldots, p_n$. Combining the relation $p \mid p_1 p_2 \ldots p_n$ with $p \mid P$, we arrive at $p \mid P - p_1 p_2 \ldots p_n$ or, equivalently, $p \mid 1$. The only positive divisor of the integer 1 is 1 itself and, since $p > 1$, a contradiction arises. Thus no finite list of primes is complete, whence the number of primes is infinite.

Appendix 5.2. Universal generalization

The rule UG

Let θx be any open sentence (or predicate) with occurrences of a variable x none of which are bound in θ. Let $\theta x/c$ be the sentence that has occurrences of the individual constant c wherever θ has x. Let $\forall x \theta x$ be the sentence expressing that everything (i.e. every member of the domain) satisfies the condition θx. Then one may infer

$$\frac{\Gamma \vdash \theta x/c}{\Gamma \vdash \forall x \theta x}$$

provided that c appears neither in θx nor in any member of Γ.

The symbol configuration says that if $\theta x/c$ is deducible from a set of sentences Γ, $\forall x \theta x$ is deducible from Γ (if the proviso is met); alternatively put, $\forall x \theta x$ is deducible from $\theta x/c$ on the assumption that all the premisses from which $\theta x/c$ is deduced are true (if the proviso is met).

The proviso is indispensable. (i) Let θx be "$x = c$", so that $\theta x/c$ is "$c = c$". $\{\forall x(x = x)\}$ entails that $c = c$, but not that $\forall x(x = c)$. (ii) Let the domain be the class of animals and let "Dx" and "Fx" mean "x is a duck" and "x flies" respectively. Let θx be Fx. Then {∀x(Dx → Fx), Dc} entails that Fc, but not that ∀xFx.

The soundness of UG

Let "model of θx/c" / "model of Γ" mean "interpretation under which θx/c is true" / "interpretation under which all sentences in Γ are true (or have true universal closures)". Suppose that any model of Γ is a model of θx/c, and let M be any model of Γ. As c does not occur in any sentence in Γ, all of them will be true under any interpretation that differs from M only in the object that it assigns to c. Thus for any object **o** in the domain of M, if M^* differs from M at most in assigning **o** to c, M^* is a model of Γ; hence M^* is a model of θx/c. Hence any interpretation that differs from M at most in what it assigns to c is a model of θx/c. Hence M itself is a model of ∀xθ.

6
Mental Number Lines

The next three chapters look at visual thinking in connection with the positive integers and their arithmetic. Visual representations are spatial representations, yet numbers, unlike geometrical figures, are not inherently spatial. So a reasonable first thought is that visual thinking in arithmetic will be insignificant and peripheral. I hope to convince you over the following three chapters that that is wrong.

The main concern of this chapter is the association of numbers with a visually represented line. The association is usually acquired in junior school, I would guess, and is then extended to provide an integrated representation of integers, fractions, and irrational numbers, called *the real number line*. This is an essential item in the toolbox of professional users of mathematics; its importance becomes clear as soon as one recalls that two or three such lines are used to form a spatial co-ordinate system for the purposes of calculus. Yet the nature and origin of mental number lines, and the way they are embedded in our thinking, are not obvious.[1] My aim is to get clearer about mental number lines given the evidence to date; I shall try to show how innate and cultural factors interact to determine the nature and role of mental number lines in basic numerical thinking; and I shall underline their importance in more advanced mathematics.

For this we first need to take a look at our (other) basic number representations. For representing positive integers we have

(1) natural language number expressions (spoken and written), and
(2) numeral systems, such as the decimal place system.

There is now considerable evidence that we also have

(3) an innate sense of cardinal size.

Although this sense is of crucial importance for numerical cognition, it is not well known. So I will now describe it and its relation to our other forms of number representation. Then it will be possible to gauge where a number–space association fits in.

Innate sense of cardinal size

This number sense is a capacity for detecting the (approximate) cardinal number of a set of perceptually given items such as a pack of predators or a sequence of howls or a bunch of bananas. The capacity is exact for very small numbers, which means that it enables us to discriminate reliably a small number from its neighbours. But for a larger set of things one can sense not its exact number but only an interval into which it falls, the larger the set the wider the interval. It is possible that there are two innate systems in operation here, one for exact representation of very small numbers and one for approximate number representation.[2] In that case the number sense should be regarded as comprising both systems. Our number sense is innately given, but it is not innately fixed. Rough number discrimination becomes more precise over the months of prelinguistic development[3] and the limit of the capacity for exact number discrimination may be pushed up. Experience with finger arithmetic,[4] verbal counting, abacus practice, and the like may sharpen the rough number sense so as to provide reliable discrimination into double-digit numbers, thus extending the range of exact cardinal number representations. But the number sense itself is an innate quantity spectrum, on a par with our sense of duration and our sense of spatial distance.

Why do we take them to be innate? Because animals and human infants have them. In some animal experiments the animal must make a certain kind of response, such as pressing a lever a certain number of times to get a reward; in other experiments the animal must discriminate between sets of presented stimuli, such as pieces of food, on the basis of the number of items in the sets.[5] Experiments on birds and non-human primates[6] and on human infants[7] show that they too can detect small cardinal numbers and discriminate between them. Experiments with rhesus monkeys showed that they can not only discriminate between sets of one to nine members on the basis of number but could also tell which of any pair of the sets is numerically greater.[8] So the monkeys could both identify a number in the range 1 to 9 and order numbers in that range by size.

Adult sense of number size

A normal child with decent education will learn to count, understand the decimal place system, acquire a store of single-digit arithmetical facts, pick up some general equational rules, and master some calculation algorithms. So, while a sense of number size is useful in the wild e.g. for rapidly gauging the number of nearby predators, in numerate civilizations one might expect it to be an unused vestige of primitive cognition. In fact that is very far from true. For a hint of the continuing importance of number sense for numerate adults, consider the following story. You ask some students if any of them can work out the value of seven to the power of six; one of them quickly writes down "1,000,000" saying that this is the answer in base 7 notation. Understanding the place system of numerals, you will see that this smart-alec answer is correct. Even so, it will probably leave you feeling somewhat in the dark. Why? It correctly designates the number, and it does so in a language you understand. Given any other number in base 7 notation you would be able to tell which of the two is larger, and the algorithms you know for multi-digit addition and multiplication work just as well in base 7 notation. So what is missing? What you lack is a sense of how large this number is. Obviously it is smaller than a million. But is it smaller than half a million, a quarter of a million, a hundred thousand, ten thousand, one thousand? It is difficult to tell without going some way towards calculating seven to the power of six in decimal notation.

The difficulty we have with this question is not because the number is large. Consider a much smaller number presented in base 2 notation: 101101. Is this smaller or bigger than forty? Again, in order to answer this you will probably have to go some way towards translating the digit string into decimal notation or natural language number expressions. Why is this? Why is it that you have a good idea of how large the decimal 45 is but a poor idea of how large the binary 101101 is? The reason is that a strong association of number size representations with decimal numerals and with your natural language number expressions has been established in your mind, while no such link has been established between number size representations and multi-digit numerals in other bases.

Further evidence that our number size sense gets mapped onto our representations of familiar numerals is indicated by a phenomenon known as the Stroop effect for numbers. The task is to indicate as fast as possible which of a pair of numerals flashed up on a screen is presented in the larger font size. Among the pairs presented are these:

[2 9]   [2 9]   [2 9]

People respond to the task faster for the second pair than the third.[9] Notice that larger font size coincides with larger number size in the second pair, while in the third pair larger font goes with smaller number. If we automatically activate representations of the cardinal numbers designated by the digits, and use those representations to judge which is the larger, that would promote the response for *nine* in these tests, thus facilitating the correct response for the font size judgement in the second test and hindering it in the third. This explains why responses for the second pair are faster than responses for the third pair. So, even when number size is irrelevant to the task at hand, presented with numerals in a familiar system we automatically access our sense of the numbers designated by those numerals and order them by size.

Further experiments reveal that even when a digit is presented too quickly for us to be aware of seeing it, our sense of the corresponding number size is accessed.[10] Yet other experiments show that automatic access of number sense is not restricted to single digits. You have a sense of the size of 45 (which in binary notation is 101101) and perhaps a vague sense of the size of 117,649 (which in base 7 notation is 1,000,000). All this attests to the fact that normal adults have a sense of number size that is not dormant. But how important is it? We best know how important some faculty is to us when we have some idea what it is like to be without it. This is revealed by the case of a bright young man studied by Brian Butterworth. This man lacks nothing but number sense and those abilities that build on it.[11] Subtraction, division, and multi-digit calculation were impossible for him, and single-digit sums and multiplications were solved slowly, using finger counting. For this man, ordinary activities such as shopping are awkward, to say the least. So it appears that we cannot acquire normal arithmetical abilities without the number sense.

Number comparison and number sense

The nature of our capacity for sensing magnitudes of one kind or another is often illuminated by comparison tasks. In number-comparison experiments subjects may be asked to indicate which of two given numbers is the larger, and the time taken to respond (or RT for "Reaction Time") is measured. An alternative is to specify a reference number beforehand, and ask subjects to indicate whether a given number is larger or smaller than the reference number. Number comparison experiments vary in the format of the given numbers (number words, arabic digits, sets of dots) and in the manner of responding.

There are two robust findings for comparison of numbers with one or two digits, the distance effect and the magnitude effect.

The distance effect: the smaller the difference between the numbers to be compared, the slower the response, for a fixed larger number. So it takes longer to respond for {6, 8} than for {2, 8}.

The magnitude effect: the larger the numbers, the slower the response, for a fixed difference. So it takes longer to respond for {9, 12} than for {2, 5}.

For single-digit number comparison the reaction time data conform pretty well to a smooth logarithmic "Welford" function:

$$RT = a + k.\log[L/(L-S)],$$

where L and S are the larger and smaller quantity respectively, and *a* and *k* are constants. Even double-digit comparison reaction times approximate to the Welford function. These phenomena are typical of response data for comparison of physical quantities that are non-discrete, such as line length, pitch, and duration.[12] This has led researchers to conclude that the mental number representations used in these tasks are quantities of a non-discrete analogue magnitude.[13] It is at this point that we hear of a mental number line: "the digital code of numbers is converted into an internal magnitude code on an analogical medium termed *number line*", says one article in a top cognitive science journal.[14]

The number comparison effects clearly do not justify the idea that number is mentally represented as line length—the same effects are found with comparison of sound volume but we are hardly tempted to talk of a mental volume line—and in fact the idea of a mental number line is often regarded as metaphorical.[15] But it is widely held that those data do justify the claim that cardinal numbers are represented by quantities of an internal analogue magnitude, where this is taken to imply that the representing magnitude is non-discrete.

MENTAL NUMBER LINES 95

This is too hasty. The reaction time data can be explained using a discrete representation of cardinal numbers: specifically, each number n is represented by n activated units, and the representation of each number includes the representation of smaller numbers.[16] Using this "discrete thermometer" model of number representation together with a certain computational model of number comparison, Marco Zorzi and Brian Butterworth found that RTs reproduced the distance and magnitude effects and conformed to a Welford function.[17] To explain the difference effect on this model, consider, for example, the pairs {6, 8} and {2, 8}. There is a difference of two nodes in the representations of 6 and 8 and a difference of six nodes in the representations of 2 and 8. This means that there is a greater difference of input activity to the response nodes for the pair {2, 8} than to the response nodes for {6, 8}, and so the competition between the response nodes for {2, 8} is resolved more quickly.

What about the magnitude effect? This is due to a feature of the decision process, namely, that the output level of a response node is a sigmoidal function of the input level, as illustrated in Figure 6.1.

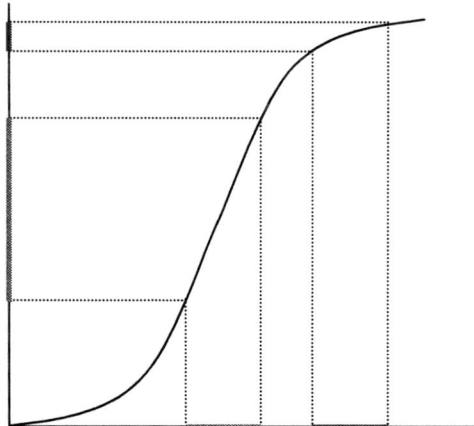

Figure 6.1

This means that outputs of nodes for numbers (above the first few) will increase with the numbers but at a falling rate; so the difference in output for the pair {3, 4} will be larger than the difference in output for the pair {8, 9} even though the input differences are the same for both pairs. Because the output difference is smaller for the pair of greater numbers, the competition between the response nodes for the greater numbers is resolved more slowly. Hence the magnitude effect. It may be that these effects could be reproduced using a similar computational model of comparison whenever the system of quantity representations is *cumulative*, in the sense that the representation of each quantity includes the representations of all smaller quantities. But the representations need not be continuous (or non-discrete) like an uninterrupted line.

What the comparison task data do rule out for the cardinal number representations of double-digit numbers is that a digit by digit algorithm is used, first comparing the left digits and then, if the left digits are the same, comparing right digits.[18] For single- and double-digit number comparison the pattern of RTs matches that for comparison of quantities such as sound volume and duration. So it looks like we use the number sense for single- and double-digit comparison.[19] For three-digit comparison, we seem to use the digit by digit algorithm; but this of course piggybacks on single-digit comparison.

What can we say about number sense at this stage? First, we have an innate rough number sense that is precise initially for numbers up to 3 and then, after finger counting has been mastered, a bit further—perhaps up to 10. But beyond a small initial segment of the numbers our number sense becomes ever less discriminate. Through verbal counting up to a hundred we probably get our rough sense of the position and size of double-digit numbers—size and position go together as the size of n is the size of the set of (positive) number expressions up to and including the expression for n in their canonical ordering. Through use of numerical symbols (numerals and verbal number terms) links are established between representations of these symbols and number sense representations. Experiences of various kinds may sharpen and strengthen our sense of double-digit number size. For example, using a 10-centimetre ruler with centimetre and millimetre marks to measure a distance to the nearest millimetre might contribute to our sense of the size of one hundred and a sense of the relative sizes of units and tens to one hundred. With the acquisition and use of the decimal place system of number notation our number sense can be mapped on to triple-digit representations and beyond. At some stage our number sense runs out, but those whose occupation involves trading in larger numbers may have a much more extended number sense than usual.

The number sense then is an innate faculty that is strengthened and refined under the impact of cultural practices. In this respect an adult's number sense is like an adult's sense of colours or sense of shapes. But there is no reason to think that the number sense consists of or depends on visual or spatial representations, or representations of some continuous spatial magnitude. In particular, nothing justifies taking the spectrum of number size representations to constitute a mental number line.

Association of number and space: the SNARC effect

But there is evidence of an *association* of number with space. In a number comparison experiment run by Stanislas Dehaene and colleagues, subjects had to classify a number as larger or smaller than 65, using response keys, one operated by the left hand and the other by the right.[20] Half of the subjects had the key for responding "smaller" in their left hand; the other half had the key for responding "smaller" in their right hand. So the two groups can be classified as (i) smaller-left and larger-right (SL) and (ii) larger-left and smaller-right (LS), as illustrated in Figure 6.2.

- 'Is the presented number *Smaller* or *Larger* than 65?'
- SL subjects responded faster (and with fewer errors) than LS subjects

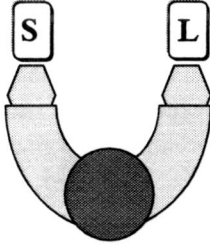 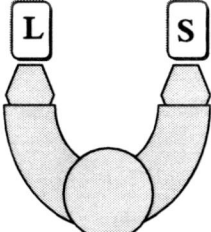

Figure 6.2. The SNARC effect
Source: Dehaene, Bossini, and Giraux (1993)

Dehaene noticed that the SL half responded faster (and with fewer errors) than the LS half. When the presented number was smaller than 65 SL subjects pressed their left-hand key faster than LS subjects pressed their right-hand key; when the presented number was larger than 65 the SL subjects pressed their right-hand key faster than LS subjects pressed their left-hand key. What could explain the reaction time superiority of SL subjects? If subjects associated smaller numbers with the left and larger with the right, correct responses would have to overcome an obstructive incongruity for LS subjects: numbers associated with the left would have to be classified by the right hand and numbers associated with the right would have to be classified by the left hand. Hence LS subjects would be slower, as was in fact the case.

But is the association with the hands? Or is it with the sides of space from the subject's viewpoint? In fact the hands are irrelevant. When subjects respond with hands crossed, subjects who have the "smaller" key on their left (but operated by their right hand) and the "larger" key on their right (but operated by their left hand) responded faster. So it is the left and right halves of egocentric space that are associated with smaller and larger numbers respectively.[21] Dehaene named this the SNARC effect.[22] Another question: What determines whether a number is regarded as small or large? It depends on whether the number falls into the lower or upper half of the test range, which subjects are made aware of prior to testing. When the range is 0 to 5, responses for 4 and 5 are made faster with the key on the right; but if the range is 4 to 9, responses for 4 and 5 are faster with the key on the left. This relativity to range excludes explanations of the SNARC effect based on properties of the digits, such as visual appearance or frequency of usage.[23]

A natural hypothesis is that for a number comparison task the number–space association is activated and the task converted into one of finding relative positions on a left to right number line. But the SNARC effect is also found in number tasks for which the size of the number is irrelevant. In one experiment subjects were asked to judge the parity (odd or even) of the presented number. For each subject the assignment of "odd" and "even" response keys to left and right was changed so that for half of the trials the "odd" key would be on the left, and for half on the right. Regardless of parity, responses to numbers in the upper half of the test range were faster when the appropriate response key was on the right, and responses to lower half numbers were faster when the appropriate response key was on the left.[24] This suggests that the number–space association is active even when it is not used to perform the current task; and that fact highlights the possibility that it is not used even in number comparison tasks, though it could be used for those tasks. Present evidence, I believe, is insufficient for a verdict on this question.

MENTAL NUMBER LINES 99

What causes this association of the left-right dimension of egocentric space with number in order of magnitude? This was investigated by using as subjects some Iranian students living in France who had initially learned to read from right to left, instead of left to right as Europeans do. Those who had lived in France for a long time showed a SNARC effect just like native French students, while recent immigrants tended to show a *reverse* SNARC effect, associating small numbers with the right and large numbers with the left. Thus all the subjects showed an association of number size with the left-right dimension of egocentric space, but the direction of the association appears to be determined by exposure to cultural factors, such as direction of reading and of ruler calibration.[25] The reverse SNARC effect has also been found in Arabic monoliterates; the same study found a weakened reverse SNARC effect in Arabic biliterates and no effect on illiterate Arabic speakers.[26] Very probably, then, this number-space association is learned, not innate. But there may very well be an innate propensity in operation here. A left-right association has been found for familiar ordered sets of non-numerical items, namely, months and letters.[27] This suggests that we have a tendency to form a linear spatial representation of any set of things whose customary presentation is well ordered (in the mathematical sense). A further indication of an innate propensity is that a small percentage of us form idiosyncratic number-space associations. Among them are calibrated curved lines with loops, strips with differently coloured intervals for different number intervals proceeding upward and rightward, complex spatial arrangements of the numerals in a combination of lines and rectangles, and many more, none of which were taught.[28]

The standard left-right number-space association can easily be overridden by another one. Daniel Bächtold and colleagues obtained a SNARC reversal within subjects, by getting them to indicate as quickly as possible whether a given number between 1 and 11 (other than 6) is larger or smaller than 6 using right and left response keys under two different conditions.[29] In the first condition subjects were led to visualize the numbers on a 12-inch ruler, and in the second condition they were led to visualize the numbers on an hour-marked clock face, though the subjects may not have been aware of using visual images; otherwise the conditions were identical. See Figure 6.3.

While on the ruler the larger numbers would be imagined on the right, on the clock face the larger numbers would be imagined on the left. Sure

- "Is the presented number *Smaller* or *Larger* than 6"

Experiment 1. Ruler display
during 20 practice trials:
"Locate the position for 5."

| | | | | | | | | | | | |

100 trials SL; 100 trials LS.
SL responses faster than LS.

Experiment 2. Clock display.
"Locate the position for 7."

LS responses faster than SL.

Figure 6.3. Reverse SNARC effect
Source: Bächtold, Baumüller, and Brugger (1998)

enough, subjects showed the SNARC effect under the first condition and those same subjects showed the reverse SNARC effect under the second condition. This points to the operation of visual imagery. In the first case the effect was probably due to the use of a visualized horizontal number *line* calibrated from left to right; in the second case the effect was probably due to a visualized number *circle* calibrated clockwise.

Association of number and space: bisection shift

The SNARC effect reveals a mental association of left-right with smaller-larger (for Western-educated subjects). Does this really justify talk of a mental number line? On its own, no. But there is further evidence for a mental number line. This comes from recent clinical data. The patients concerned suffer from a visual deficit known as neglect.[30] Neglect patients fail to notice objects and events on one side of their visual field, usually the left, following brain lesions on the opposite side usually of the inferior parietal lobe. (Occasionally the deficit relates to objects rather than to the whole visual field, so that the left side of an object anywhere in the visual field will go unnoticed.) Neglect is not the same as hemianopia (left-field or right-field blindness), as there is plenty of evidence of visual processing on the affected side.[31] Rather, it is usually regarded as a loss of visuo-spatial attentional control, an inability to attend to features on one side of the visual field resulting in a loss of visual awareness on that side. Neglect patients may leave the food uneaten on the left side of the plate, may shave only the right side of their face, or miss words on the left side of a page when reading.[32] When asked to draw a copy of a picture presented to them, e.g. of a cat, they may draw just the right half; if the picture is of a clock face they may omit the numerals on the left side.

These symptoms reveal a deficit of perception. A parallel deficit of imagination has been found to accompany it. A remarkable example of neglect in visual imagery was provided by two neglect patients from Milan.[33] They were asked to visualize and describe Milan's Piazza del Duomo from the side of the square facing the cathedral. Both patients described features that would have been on their right from that viewpoint but omitted those that would have been on their left. Afterwards they were asked to visualize and describe the square from the opposite side, as if they were standing just in front of the cathedral facing away from it. Then they described features that were previously omitted, and they omitted features previously described; so they were reporting just those features that were on their right in the currently imagined scene and on their left in the previously imagined scene.

A symptom of neglect relevant here is that when asked to mark the midpoint of a horizontally presented line segment, patients typically choose a point to the right of the actual midpoint, as in Figure 6.4.

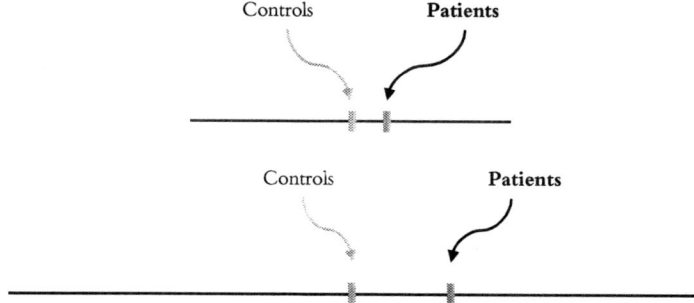

Figure 6.4

For a given line, the extent of shift to the right varies among patients. For a given patient, longer lines mean greater rightward shift.[34] Corresponding to the line bisection test there is a number-range "bisection" test: subjects are presented with two numbers and are asked to state the number midway between two given numbers, e.g. 5 for {3, 7}, without calculation. In a recent study Zorzi and colleagues reasoned that if the mental number line were a fiction, neglect patients performing number bisection would not show a shift above the mid-number corresponding to the rightward shift shown in line bisection.[35] They tested four patients with left-side neglect, four right-brain damaged patients without neglect, and four healthy subjects, all with normal numerical and arithmetical abilities on standard tests. While the healthy subjects and non-neglect patients showed no deviation from the mid-number in the number bisection tasks, the neglect patients systematically overshot the mid-number. Moreover, this shift above the mid-number almost always increased with the range (i.e., the difference between the given numbers), thus replicating the pattern of line bisection errors typical of neglect patients.

Although this is not an overtly visuo-spatial task, the data can be explained on the assumption that number representations are integrated with a visually imagined horizontal line, or a horizontal row of evenly spaced numerals, and we attempt to locate the mid-number of the given range by means of an unconscious internal line bisection, choosing the number represented closest to the bisection point. When patients are given a pair of numbers, say 2 and 9, with the task of choosing a mid-number, an image of the segment of the number line from 2 to 9 is activated automatically and unconsciously. There is no loss of the leftward part of the line-segment image because attention is not required to produce the image. But attention is required to *use* the image, even when the use is not conscious; so in left-neglect patients the leftward part of the image is not available for use.

There are a couple of worries one might have about the hypothesis that number bisection involves the use of a visual number line. One arises from the fact that the subject has no consciousness of using a number line in performing number bisection. How can an image, a visual image, occur without the subject's being conscious of it? To a philosopher the notion of an unconscious image is liable to sound incoherent. The solution is to distinguish between the *phenomenal* image, that is, an item subjectively present in the conscious sensory experience of the subject, and what I will call the *psychological* image, which is a pattern of activity in a specialized visual buffer generated solely by top-down processes.[36] When such activity is very weak or very brief, it may suffice to affect reaction times but be insufficient for the occurrence of an item in the conscious experience of the subject: there will be a psychological image without a phenomenal image.[37]

The second worry concerns the role of attention. On the explanation proposed here, visual attention to the line-segment image is involved in the patients' performance in the number bisection task (and errors are ascribed to a defect of visual attention). If so, there is attention to the line-segment image without any awareness of it. But how can there be attention without awareness? Attention in this case is not directed by the subject; the relevant deployment of attention is an automatic sub-personal process, a selective increase in activation that makes the information of the selected part available for further processing. The reaction times in bisection (and comparison) tasks are fractions of a second; the components of such rapid processes, including those that constitute shifting or focusing attention, may easily be too brief for the subject to be aware of them.

More recent data, however, weigh against the hypothesis that number bisection involves the use of an unconscious horizontal number line in the visual imagery system. Doricchi and colleagues have found a double dissociation between rightward shift in line bisection and upward shift in number interval bisection.[38] By mapping and comparing the lesions in neglect subjects who showed upward shift in number bisection and those who did not, it was found that the damage probably responsible for the upward shift lay outside the areas most frequently damaged in neglect subjects. Moreover, brain imaging studies suggest that different cerebral regions are activated in number comparison tasks and visual line bisection. At this stage, then, it is likely that if the number–space association revealed in the standard SNARC effect data is used in number bisection, it is independent of any visual number line representation.

Calibration

The expression "mental number line" may be applied to three things:

(1) the sequence of number magnitude representations of the number sense;
(2) the association of number sense representations with positions in egocentric space;
(3) the association of number sense representations with positions on a visual line.

In each case there arises a question about the association, which we may think of as a question of calibration. Is it regular as on an ordinary ruler?

Or increasingly compressed as on a slide rule? What exactly this comes to varies in each case; so let us look at each case in turn.

In the first case, consider two pairs of numbers, each pair differing by the same amount, such as {2, 5} and {9, 12}. It is quite possible that although the *numerical* difference is the same for both pairs, the *cognitive* difference between the number sense representations for 9 and 12 is smaller than the *cognitive* difference between the number sense representations for 2 and 5. The cognitive difference between two number sense representations is an explanatory factor in performance of cognitive tasks, a factor that can be described as the subjectively apparent difference. The increasing compression hypothesis for the sequence of number sense representations is that this holds in general: for a fixed numerical difference, the greater the mean of a pair of numbers, the smaller the cognitive difference between their number sense representations. Another way of putting it is that the number equal to $n \times k$ seems (to the number sense) less than n times as large as k.

Why would anyone hold this increasing compression hypothesis for the number sense? One reason is that it explains the magnitude effect in number comparison tasks: when comparing two numbers, the larger the numbers, the slower the response, for a fixed difference. So it takes longer to respond for {9, 12} than for {2, 5}. This is just what would be expected if the cognitive difference between the number sense representations of 2 and 5 is larger than the cognitive difference between the number sense representations of 9 and 12.

However, there is a rival explanation of the magnitude effect. On the rival view, the association between numbers and number sense representations is not invariant. Perception of an array of n things, or of a corresponding numeral, will not always activate the same number sense representation; on the contrary, over time there will be a distribution of number sense representations activated by perception of a given numeral, and the standard deviation of this distribution increases in proportion to the number denoted by the numeral. This is known as the scalar variability hypothesis, as variability increases to scale. Here then is the rival explanation of the magnitude effect: the ease of discriminating two numbers falls as the numbers get larger, not because corresponding number sense representations become more similar, but because the variability increases. This means that the mapping of numbers (presented as numerals or sets of perceived objects) to number sense representations gets noisier for larger numbers. More specifically, the overlap of distributions of number sense representations to which "9" and "12" get mapped will be greater than the overlap of distributions of representations to which "2" and "5" get mapped.

Gallistel and Gelman give reasons to prefer the scalar variability hypothesis to the increasing compression hypothesis.[39] One reason is that the psychophysics of number and duration discrimination appear to be identical in animals, and it has been shown that the mental represention of duration is a linear function of actual duration, rather than a logarithmic (or other compressive) function of actual duration. Another reason is that if the mapping of numerals to number sense representations is logarithmic (or otherwise compressive), plausible modelling of the cognitive operation of rough addition is well-nigh impossible, whereas if the mapping is linear, plausible modelling of rough operations of addition and multiplication are available.

But Dehaene cites other results as reasons for preferring the increasing compression hypothesis.[40] When subjects are asked to produce random numbers in a given interval, they typically produce more small numbers than large numbers.[41] In a related experiment, subjects are presented with a series of numbers in a given interval (not in order of numerical size) and asked to judge how evenly and randomly the numbers in the series are drawn from the interval.[42] Here are a couple of such series drawn from integers between 1 and 2000. Which one seems to you the more random and the more evenly spread?

A: 879   5   1,322   1,987   212   1,776   1,561   437   1,098   663
B: 238   5   689   1,987   16   1,446   1,018   58   421   117

Most people find series **B** the more random and more even sample; in series **A**, large numbers seem over-represented. In fact, series **A** is the more evenly spread, with intervals of just over 200. In **B**, the intervals decrease exponentially.

This does not seem to me to be strong evidence in favour of the compression hypothesis. Smaller numbers may be over-represented in our "mental urn", as Dehaene puts it, because of our greater use of and exposure to smaller numbers. Another possibility is that we are unconsciously categorizing the numbers by digit length (number of digits in the numeral), thereby giving equal weight to the set of 9 single digit numbers, the set of

90 double digit numbers, the set of 900 three digit numbers, and the set of 9, 000 four digit numbers. With that weighting, series **B** is indeed more evenly spread.

Dehaene also mentions a small informal number bisection test reported by Attneave. Fourteen adults were asked for a quick intuitive answer to the following: suppose that *one* is a very small number and *a million* is a very large number; now give a good example of a medium-size number. Though 500,000 is midway, the median of the "bisections" obtained was 100,000.[43] On the assumption that the number sense representations were used in this task, this underestimation is what would be expected on the increasing compression hypothesis. But there is an alternative explanation, which dispenses with the somewhat implausible assumption that number sense representations are being used for such a large range of numbers. This focuses on linguistic salience among the ordinary verbal number expressions. Powers of ten are more salient than intervening numbers, and so 100,000 is likely to be the largest mental pole of attraction short of one million. Sophie Scott and colleagues give indirect evidence of this.[44] Perhaps that is why the median of responses to Attneave's test was 100,000.

Dehaene cites a third piece of evidence as decisively favouring the increasing compression hypothesis.[45] Neurons have been found in the primate brain whose firing rate is tuned to specific cardinal numbers. A neuron for 3 would respond optimally to a display of three visual objects, less to two or four objects, and not at all to one or five objects. Nieder and Miller analysed the response curves of number neurons in two monkeys engaged in discriminating the cardinality of two visually presented sets.[46] Dehaene reports that these neural tuning curves, when plotted on a linear scale, are asymmetrical and assume different widths for each number (in the range 1 to 5), but "became simpler when plotted on a logarithmic scale: they were fitted by a Gaussian with a fixed variance across the entire range of numbers tested. Thus, the neural code for number can be described in a more parsimonious way by a logarithmic scale than by a linear scale." Without some further assumption this is hardly decisive. The fact that curves look simpler when the data are plotted one way rather than another is not enough to decide between competing hypotheses. As a criterion of theory choice, parsimony must be applied globally to explanatory theories or models in the context of relevant background knowledge, rather than to the presentation of data sets. It may in the future turn out that, all things considered, the most parsimonious explanation of all the relevant data implies a logarithmic scale. But at present the question seems to me undecided.

What about the calibration of the two other kinds of mental number line, the association of number representations with positions in egocentric space, and the association of number representations with positions on a visual line? I do not know of data that support the hypothesis that the number–space association is increasingly compressed. If the number–space association is in play in number bisection tasks, we should expect to find normal subjects showing a leftward shift, one which increases systematically with the size of the number interval, on the increasing compression hypothesis. For example, if log to the base 2 is the function which describes the mapping of number representations to spatial positions, each successive unit of spatial distance would have twice as many number representation mapped onto it as its predecessor: if 2 is placed at one unit along, 4 will be placed at two units along, 8 at three units along, 16 at four units along, 32 at five units along, and 64 at six units along. So 4 would seem subjectively half of 16, and 8 would seem subjectively half of 64. But when we look at the performance of controls in the number bisection tasks in the studies by Zorzi and by Doricchi and their colleagues mentioned earlier, we find no systematic leftward shift increasing with the number interval. Conditional on the assumption that number bisection involves the number–space association, this would be evidence against the increasing compression hypothesis for the number–space association.

What about the calibration of the mapping of number sense representations to positions on a visual line? The full answer to this depends on the nature of the visual number line, which is the topic of the next section; calibration of the visual number line will be considered in that context.

The nature of a visual number line

We typically visualize a number line as a graphical line with numbers represented as positions on the line ordered from left to right for individuals in Western cultures. But there are many possible variations. Are the number positions marked in the image, say with little vertical line segments across the horizontal line? Does an image of the corresponding arabic numeral appear just above (or below) each number position? Or are just some number positions, say multiples of 5, thus labelled, or none at all? Probably such details vary from person to person, and perhaps from one occasion to another for each of us. What seems likely to be constant is that each number is represented by a position on the line (or in the row of numerals) relative to a unique *origin*, i.e. a left (or right) end or a zero-marked position for lines (or rows) extending endlessly in both directions, and that the size of the number is represented by the relative distance between the origin and the number position. A left-ended line can be extended endlessly leftwards to accommodate negative numbers, but a line that accommodates both positive and negative numbers appears to be a separate representation from the standard left-ended line.[47]

The infinity of a visual number line   But a visual image of a line that is endless in one or both directions, an infinitely extended image, is surely impossible. The field of visual imagery is surely bounded. Imagine walking towards an adult giraffe from the side; the visualized giraffe will seem to loom larger as you continue the mental walk until not all of it can be visualized simultaneously, head and hoofs beginning to "overflow" image capacity. This phenomenon has been tested and confirmed: image size is constrained.[48] So there will be an upper bound on initial segments of the number line that we can visualize when number marks appear clearly and evenly separated by a fixed distance. But when we talk and think of the number line, what we have in mind is an infinitely extended line.

To best resolve this problem we should make a distinction between two kinds of representation used by the visual system. Here I follow Kosslyn's account,[49] but any account of the visual system will need the following distinction. There are on the one hand category specifications, and on the other hand visual images. A category specification is a set of related feature descriptions stored more or less permanently; the category specification for squares given in Chapter 2 is an example. A visual image is a fleeting pattern of activity in a specialized visual buffer, produced by activation of a stored category specification.[50] The category specification can specify that a line continue in a certain direction endlessly; but for a single momentary image generated by activation of that category specification only a finite segment of a line will be represented in the image.

From a given category specification more than one image can be generated; in fact a sequence of continuously deforming images can be generated over an interval of time that we can think of as a single continuously changing image. When the category specification is activated, its feature descriptions become input for a system that I will call the image generating function.[51] That function has additional "parameter" input variables corresponding to viewpoint, distance, orientation and perhaps others. These variables can be continuously changed and when that occurs the result will be continuously changing imagery. Imagine a regular mug on an eye-level shelf where it is stored upside down with its handle to your right. Now imagine it as you take it down and bring it to a position and orientation that allows you to look into it from above. In that case the image generating function acts on the category specification for a regular mug with continuously changing parameters for viewpoint, distance, and orientation. Figure 6.5 is a diagram, modified from Kosslyn, of a part of the visual system relevant here, omitting all arrows other than those involved in generating and transforming images. The vertical downward arrows to the box for the image generating function indicate the inputs that constitute parameter values.

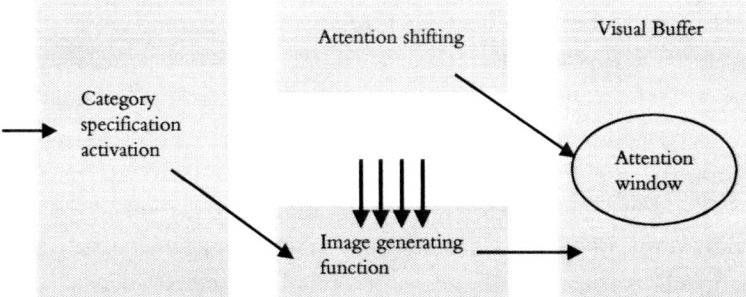

Figure 6.5. *Source*: Adapted from Kosslyn (1994)

Among the image-transforming operations resulting from continuous changes of parameter inputs to the image generating function there are scanning, zooming (in or out), and rotating.[52] For example, if we want to "locate" a large number, say 232, we might "scan" the line until the image includes a portion that covers the 230s and neighbouring decades, then zoom-in on just the decade 230–9, and finally shift the attention window to the position of 232.[53] Of course, these imagery operations of scanning and zooming-in cannot be what they are in visual perception. In particular, they are not operations on a fixed image. Imagistic zooming-in is transforming the image by smoothly changing one or more egocentric distance parameter values for the image generating function. This continuous change of image has the subjective visual effect of moving towards the imaged object, thereby increasing resolution. Imagistic scanning, like imagistic zooming-in, results from smooth change of a parameter value for position.

How does all this relate to the infinity of the number line? The number line category specification contains a feature description to the effect that the line has no right end; this ensures that rightward imagistic scanning will never produce an image of a right end-stopped line. This is what constitutes the infinite extension of the number line in the visual imagery system. There is no implication in this account that a visual image can be infinite in extent.

The calibration of a visual number line   Now let us return to the question whether the calibration of a visual number line is regular or increasingly compressed. My conjecture is that the category pattern specification for a number line entails that the number line has regular calibration. But not all of the number line *images* produced by activation of the category pattern have regular *apparent* calibration. Some images will have apparent calibration that matches a perspectival projection; so the calibration will appear increasingly compressed. To see how this is possible, recall that a current visual image depends not only on the activated category pattern but also on parameter inputs to the image generating function. One of these will determine whether the image viewpoint is perpendicular to the number line or oblique. If perpendicular, the calibration appears regular in the image; if oblique, the calibration appears increasingly compressed in the image. Which of these kinds of viewpoint (perpendicular or oblique) is selected may depend, at least in part, on the size of the numerical interval set by the task at hand. For small number intervals, a perpendicular viewpoint is likely to be selected, dictating regular image calibration. For large number intervals, the imagery system could first zoom-out and then rotate, thereby selecting an oblique viewpoint; the result would be an image in which the calibration appears increasingly compressed.

MENTAL NUMBER LINES   111

Conscious uses of number lines

Visual number lines are not consciously used in basic number tasks such as number bisection, but they are consciously used in other ways. It is easy to underestimate the importance of visual number lines in our mathematical thinking. First and foremost there is the use of a visual number line in integrating our grasp of various number systems and operations on them. Secondly, we have the use of orthogonal number lines in a co-ordinate system for space of two or three dimensions. Thirdly, we have the use of a number line in structural discoveries in the theory of integers. To bolster my claim that these are not trivial uses, here are some reminders.

Integrating number systems   A number line helps us first to integrate "counting" numbers with "measuring" numbers. Once we have a concept for unit length a whole number n can be thought of as the length of an initial segment composed of n adjacent segments of unit length. It can also be thought of as the position of the right end of that initial segment. We use both of these representations, sometimes in combination. Whole number addition now has an easy representation as a movement to the right from the position marking one addend by the length representing the other addend, the result being represented by the end position (or the length of the segment from the origin to that position). Whole number subtraction $n - k$ can be represented as a leftward movement from the position representing n by the length representing k, the result being represented by the end position. Allowing the number line to be extended ever leftward from the origin enables us to model subtraction $n - k$ when $k \geq n$.[54] We also have representations of multiplication, division, and rational numbers in terms of the number line — I omit the details.

So one important use of a number line is to help integrate our grasp of the system of positive integers with its extension to zero and the negative integers, and this with the system of rational numbers. It is true that we also use other ways of introducing rational numbers (pizza fractions), negatives and zero (debit and zero balance). But it is difficult to introduce real numbers beyond the rationals without using line lengths relative to some unit length.

How is it done using line lengths? We know that the positive square root function applied to 2 does not yield a rational number. What convinces us that it yields anything? For a domain that does not yet include complex numbers, we rightly think that negative numbers have no square roots; we could reasonably suspect that the same is true of positive integers which are not squares of other integers. But we know 2 does have a square root, as $\sqrt{2}$ is the ratio of the length of a square's diagonal to its edge length. That is known by thinking of the kind set out famously in Plato's *Meno* and discussed in Chapter 4. From Pythagoras' theorem we know that, for a chosen unit length, if a right-angled triangle has perpendicular sides of lengths 1 and $\sqrt{n}$, its hypotenuse has length $\sqrt{(n+1)}$. So for every positive integer $k > 1$, $\sqrt{k}$ is the ratio of the length of a certain line segment to the unit length; but unless k is the square of some smaller integer, $\sqrt{k}$ is irrational. One very natural way of understanding the real numbers takes over this way of making sense of irrational square roots of integers: a real number is the ratio of the length of a line segment to a chosen unit length. A generalization of this is in fact Newton's way of understanding the real numbers.[55] The understanding of real numbers in terms of infinite sequences (or sets) of rational numbers comes much later in the history of mathematics.[56] In the development of an individual's mathematical education it is often a matter of making sense of infinite decimal expansions, and this can be done by thinking of them as the result of a process of extending a line by successively smaller increments, a digit n in the kth place signifying an increment of $n \times 10^{-k}$ times the unit length.

Thinking in terms of a number line, in the ways just described, provides us with an integrated grasp of the number systems we use most: the integers, the rational numbers, and the real numbers. Once we have a category specification for the real number line, operations of image copying and rotation enable us to visualize two such lines perpendicular to each other and crossing at their zeros, giving a co-ordinate system for a plane. As is well known, this gives us the resources for a model of the complex number system that embeds the other number systems. No one who spends any time working with complex numbers will doubt the utility of this integrated spatial representation of the complex and other numbers systems.

Grasping qualitative features of functions on the real numbers Numbers aside, a co-ordinate system provides a means of grasping *qualitative* features of real valued functions of a real variable, thereby opening a door to real

GIAQUINTO, M. (MARCUS)

VISUAL THINKING IN MATHEMATICS: AN
EPISTEMOLOGICAL STUDY.

OXFORD: OXFORD UNIVERSITY PRESS, 2007   Cloth   287 P.

AUTH: UNIVERSITY COLLEGE LONDON.

LCCN 2007-015473
ISBN 0199285942   LIB PO#   QA8.4

181 UNIV OF MAINE/ORONO   APPROVALS
9/05/07 OSTA   5718-11

LIST	72.00
DISC	18.0%
NET	59.04

YBP - CONTOOCOOK, N.H. 03229

SUBJ: 1. MATHEMATICS---PHILOSOPHY. 2. VISUAL
PERCEPTION. 3. GEOMETRY.

CLASS QA8.4   DEWEY# 510.   LEVEL ADV-AC

analysis. An example of this was given earlier: according to the "discrete thermometer" model of number representation, the output activity level of response nodes for number representations is a *sigmoidal* function of the numerical input. Figure 6.1 illustrates this. Seeing a diagram of a sigmoidal function or visualizing one makes it easy to see that the difference in output for a pair of large numbers a fixed distance apart will be smaller than the difference in output for a pair of middling numbers the same distance apart. The conclusion that the output difference for a large pair will be smaller than that for a middling pair (when the pairs are equally distant) is, of course, not literally perceived to be true. Rather it is arrived at by inference from features of the perceived or visualized diagram together with contextually supplied assumptions about the ways in which the diagram carries information, that is, about how to "read" the diagram. I claim that this is (or can be) a reliable hence knowledge-yielding way of reaching the belief. Although the previous chapters give some clue as to how this is possible, I will not attempt to substantiate the claim here. Instead, a later chapter will be devoted to the epistemology of visual thinking in connection with functions of real numbers, as this topic is not at all straightforward and needs extended discussion.

Grasping structural facts about the integers   The fecundity of the visual device of orthogonal axes for mathematics cannot be overestimated. It gives us a basic resource for analytic geometry and more generally for real and complex analysis. What is less often noted is that the number line is sometimes useful even in the theory of integers (usually known as number theory). Thinking in terms of the number line can make a truth of number theory obvious prior to finding a proof. This can be useful both as an aide mémoire and in helping one to see whether an attempt to construct a proof is a fool's errand. Here is an illustration using a basic truth of number theory known (oddly) as "the division algorithm":

> Given any integer a and any positive integer b, there exist unique integers q and r, with $0 \leq r < b$, such that $a = qb + r$.

How might one arrive at this? First one can think of the sequence of integers as points on a line with the multiples of b marked on it. Particular numbers are not included in the image, which is an image of an arbitrary interval of the line containing several multiples of b, as illustrated in Figure 6.6.

114 MENTAL NUMBER LINES

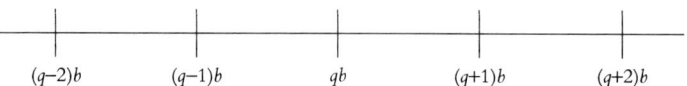

Figure 6.6

Using this image one can tell that there are just two possibilities for the position of a: either it coincides with the position of one of the multiples of b, or it falls in between two consecutive multiples of b. In the first case let q be the multiplier for which $qb = a$ and let r be 0. Then trivially $0 \leq r < b$ and $a = qb + r$, as the theorem says.

In the second case let q be the multiplier in the multiple of b just left of a and let r be $a - qb$, the distance from qb to a, as in Figure 6.7.

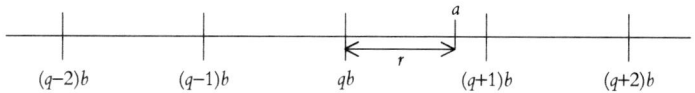

Figure 6.7

Then the figure shows again that $0 \leq r < b$ and $a = qb + r$. So either way there is a unique integer q and a unique non-negative integer r less than b such that $a = qb + r$, as the theorem says.

This visual argument is persuasive and makes the correctness of the proposition obvious in a direct way that cannot be matched by a number-theoretic proof of it. A proof would have to deploy some general principles of number theory including the least-number principle or some equivalent, such as natural number induction. The visual argument accommodates generality without using the least-number principle or any equivalent. It focuses on a general structural feature of the sequence of integers, using a number line representation that includes letter-variables to range over the integers rather than representations of particular numbers, so that there is no restriction to a particular segment of the sequence of integers. In this instance we use the fact that a positive integer has a twofold representation on the number line, as a point-position on the line and as the distance on the line between its position and the origin. With respect to the integer a here, only its representation as a position is used; with respect to the integer r only its representation as a distance is used; with respect to the integer b both its distance representation and the positional representations of its multiples are used. The multiples of a positive integer b are represented by equi-spaced positions, giving Figure 6.6; that allows us to perceive that when the position of a number *a* does not coincide with that of a multiple of *b*, there is a unique closest multiple of *b* positioned to its left. The epistemic work here is achieved by deploying one's implicit grasp of these facts of representation together with vision or visual imagination and some simple deduction. I take it that the reliability of the use of these resources illustrated here is uncontentious. I am not, however, claiming that this visual argument is a proof.

At the conscious level, then, a horizontal number line has immense value as a way of developing an integrated way of thinking of the various kinds of numbers that we make use of beyond pure mathematics; then there is the use of number lines for co-ordinate systems, and as a means of representing structural features of the integer system.

Conclusion

Sometimes I have slipped into talking of *the* number line, but I have not meant to imply that there is only one. In fact we have a satchel of representations that come under this title: a single horizontal line with one endpoint for the non-negative integers; the axes of a non-negative co-ordinate system; a horizontal line endless in both directions for the real numbers; the axes of a co-ordinate system in two dimensions; the same for three dimensions. There are also idiosyncratic number lines mentioned earlier, but their role in the mathematical thinking of their owners may be negligible. In contrast, the importance of the widely shared number lines that I have been discussing is beyond serious dispute. The conscious uses illustrated in the previous section may be too familiar to be given due credit in our mathematical thought and understanding.

The standard horizontal number line is clearly a product of culture, as it depends on culture-specific conventions of a written numeral system. But it also depends on three innate faculties: our number sense, our sense of the space around us, and the visual imagery system. The representations of our number sense are mapped onto the representations of our numerals, which in turn are mapped onto a horizontal line (or are arranged in a row) to form an integrated system of numerical representation in the imagery system; when activated, the resulting image is integrated into the representation of egocentric space. So here we have an example of a basic resource of human intelligence that is the product of an interaction between cultural and innate endowments.

Is this interaction just a lucky accident? Probably not. The fact that we so easily acquire and internalize a mental number line, as well as the fact that some of us form a mental number line independently of school instruction, suggests that we have an innate propensity to form a number line representation once we have acquired a written numeral system. Such a propensity may be a special case of a disposition to represent totally ordered systems of items, such as alphabets, as a line. It is also an instance of the disposition found in innovative mathematicians to integrate symbolic and diagrammatic representations, a disposition whose fruitfulness is beyond dispute.

Notes

1. When cognitive scientists talk of a mental number line they are often *not* talking about an acquired spatial representation. For example, the subject of a recent article headed 'The neural basis of the Weber-Fechner law: a logarithmic mental number line' is the innate system of cardinal size representation found in humans and many other animals. Dehaene (2003).
2. Xu (2003). The approximate system would probably be inoperative for sets of just one or two things.
3. Lipton and Spelke (2003).
4. See Butterworth (1999: ch. 5).
5. But can we be confident that it is really number that the animal is responding to and not some other kind of quantity? In one set of experiments a rat had to press one lever a set number of times consecutively and then press another lever once in order to get a reward. The required number of presses was varied (4, 8, 12, 16). See Mechner (1958). The rats do pretty well at this. But they could have succeeded by repeatedly pressing until a certain period of time had passed before changing levers, thereby using *duration* and not *number* as the criterion for changing levers. Further tests were done to screen this out. See Mechner and Guevrekian (1962). Moreover, experiments of a different kind showed that the brain of a rat hearing a sequence

of tones will register both duration of the whole sequence and number of individual tones. See Meck and Church (1983).
6. Gallistel (1990) gives an account of much of this research. Thorpe (1963) reports experiments on birds by Koehler. For later work on primates see Boysen (1993), and Rumbaugh and Washburn (1993).
7. Xu and Spelke (2000); Starkey et al. (1990); van Loosbroek and Smitsman (1990).
8. Brannon and Terrace (1998).
9. Foltz et al. (1984). The original experiments of J. Stroop in 1935 showed that we automatically access the meaning of a graphically presented word.
10. Dehaene et al. (1998).
11. This is the case of "Charles" reported in Butterworth (1999: ch. 6).
12. Welford (1960). Moyer and Landauer (1967, 1973) first showed that the RT data for number comparison fit the Welford equation pretty well.
13. See e.g. Moyer and Landauer (1967).
14. Dehaene et al. (1990).
15. Dehaene (1997) calls it "a simple yet remarkably powerful metaphor".
16. Zorzi and Butterworth (1999).
17. Ibid.
18. Hinrichs et al. (1981), Dehaene et al. (1990). But for three-digit numbers it seems we do use digit-by-digit comparison.
19. But there is some evidence that an additional mechanism is operative, by which tens and units are evaluated separately; however, the effect is relatively insignificant. Nuerk et al. (2001). Consider pairs that are closely matched for difference and size of larger number. A pair whose larger number has both a larger tens digit and a larger units digit than the smaller number is said to be *compatible*; a pair whose larger number has a larger tens digit but smaller units digit is said to be *incompatible*. We are slightly faster for compatible pairs than incompatible pairs. For a review of work on this question see Nuerk and Willmes (2005).
20. Dehaene et al. (1990); Dehaene et al. (1993).
21. By using saccadic instead of manual responses, Schwarz and Keus (2004) confirm that the effect is not just an overlearned motor association between manual processes and numbers (from e.g. computer keyboard

use). They also found a spatial association of large and small numbers with high and low.
22. This name, which is an acronym for "Spatial-Numerical Association of Response Codes", is a deliberate allusion to Lewis Carroll's poem "The Hunting of the Snark". See Dehaene (1993, 1997).
23. Dehaene and Mehler (1992).
24. Dehaene et al. (1993).
25. As far as I know, subjects who read downwards from the top of the page have not yet been tested.
26. Zebian (2005).
27. Gevers et al. (2003)
28. See Galton (1880) for the original studies. For excellent further investigation and discussion see Seron et al. (1992).
29. Bächtold et al. (1998).
30. It is also known as unilateral neglect or hemispatial neglect or hemineglect.
31. Cohen et al. (1995); Driver and Mattingley (1998); Schweinberger and Steif (2001).
32. Robertson and Marshall (1993). Neglect patients are usually unaware of the deficit, unlike hemianopia patients, and so do not try to compensate by motions of head or eyes.
33. Bisiach and Luzzatti (1978).
34. For example Halligan and Marshall (1988).
35. Zorzi et al. (2002)
36. Here I distinguish between visual image and visual percept, the latter being a pattern of activity in the visual buffer generated by a combination of bottom-up and top-down processes.
37. There is no concession to dualism here. The neural activity and the visual imagining are one and the same event; but this event has some properties that are not subjectively accessible (the neural activity pattern) and some that are only subjectively accessible (the phenomenal character).
38. Doricchi et al. (2005).
39. Gallistel and Gelman (1993).
40. Dehaene (1993, 1997)
41. Baird and Noma (1975).
42. Banks and Coleman (1981).

MENTAL NUMBER LINES 119

43. The mean was a little over 186,000. The standard deviation was not given. Attneave (1962).
44. Scott et al. (2001).
45. Dehaene (2003).
46. Neider and Miller (2003).
47. Fischer and Rottmann (2005).
48. Kosslyn (1978). Subjects had to visualize different sized objects one at a time, imagine walking towards the object until apparent overflow, and then estimate their apparent distance from the object. It turns out that the larger the object, the further away it seems at the point of overflow. More precisely, the estimated distance at the point of apparent overflow matches the distance at which the object subtends a visual angle just too large for its edges to be simultaneously visible.
49. Kosslyn (1994).
50. Kosslyn uses the expression "category pattern" for category specification. I changed the term to avoid the suggestion that a category specification is a kind of template as opposed to a set of feature codes.
51. Kosslyn refers to this as the "mapping function" from the pattern activation system to the visual buffer.
52. Mental rotation was discovered by Shepard and Metzler (1971). For its role in visual imagery see Kosslyn (1994). Scanning and zooming-in were investigated by Kosslyn. See the accounts of this work in Kosslyn (1980) and his theoretical account of these transformations in Kosslyn (1994).
53. Shifting the attention window to a certain part of the visual buffer increases the level of activity in that part, thereby raising the likelihood that the information of that part will be further processed.
54. Thus we can acquire a way of thinking of 0 and the negative integers that goes beyond the merely formal: 0 is represented by the position of what was the left end of the original number line, and for positive k, $-k$ is represented by the position of $0 - k$. Adding can be extended to cover all pairs of integers by taking, for positive k, $i + (-k) = i - k$.
55. "By number we understand not so much a multitude of units, as the abstracted ratio of any quantity of any kind, to another quantity of the same kind which we take as unit. And this is threefold: integer, fracted and surd. An integer is what is measured by the unit, a fraction that

which a submultiple part of the unit measures, and a surd to which the unit is incommensurable." From Raphson's translation of Newton's *Arithmetica Universalis* (1720).

56. These are 19th-cent. constructions of Weierstarss, Dedekind, Cantor and others.

7
Visual Aspects of Calculation

What kinds of thinking occur in arithmetical calculation? Is calculation *sui generis*, with peripheral use of linguistic, visual, and motor abilities? Is linguistic processing at the core of calculation, with peripheral use of visual and motor abilities? The main concern of this chapter is to describe the roles of visual thinking in calculation, with the aim of showing that in many cases it is not peripheral. I try to do that by setting the visual aspects of calculation within an account of the operations involved and the cognitive resources used. Along the way I will also consider the epistemology of basic arithmetic.

Finger-counting and single-digit addition

For almost all of us arithmetic begins on the fingers. Once we learn to count our fingers become our most readily available counters, and operations on our fingers provide our basic experiences of adding numbers. In fact finger methods constitute our chief means of coming to grasp what numerical addition is and coming to know the numerical values of single-digit sums. There are cultural differences in methods of finger counting and adding, and variations within cultures. But some use of finger counting is almost universal, and typically children will spontaneously develop finger methods of adding instead of continuing with the first method they learned; this development is often unaffected by classroom instruction and sometimes actively opposed by teachers.[1] Here is an example of a part of a developmental sequence.[2] Let the task be to add 4 and 3.

[1] *Sequence Count All*:
Count and raise in turn four fingers, keeping other fingers down ⇒ count and raise in turn three of the fingers kept down, leaving others

unchanged ⇒ count the raised fingers ⇒ take the result of this last count as the answer.

[2] *Pattern Count All*:
Make the four-finger pattern on one hand ⇒ make the three-finger pattern on the other hand ⇒ count the raised fingers ⇒ take the result of this count as the answer.

[3] *Pattern Count On*:
Make the four-finger pattern on one hand ⇒ make the three-finger pattern on the other hand ⇒ starting with the first number word after "four", count the raised fingers on the hand making the three-finger pattern ⇒ take the result of this count as the answer.

[4] *Sequence Count On*:
Starting with the first number word after "four", simultaneously raise and count in turn three fingers ⇒ take the result of this count as the answer.

Sequence counting-on allows one to find all the single-digit sums. For example 9 + 9 can be found by raising in turn and counting 9 fingers, starting the count with "ten" and counting on. An alternative is to break up a large digit to get more familiar sums, e.g. 9 + 1 + 8. Eventually we get to know all the single-digit sums; that is, we build up a store of single-digit addition facts, on the basis of our experience with finger adding.

In typical cases one gets one's store of single-digit addition facts by recruiting a combination of visual, motor, and speech skills which constitute one or other method of adding on one's fingers. Thus finger counting underlies our knowledge of single-digit sums. This fact appears related to a neural fact. Retrieval of single-digit sums from memory employs a network involving circuits in the left precentral gyrus in the proximity of the area activated in finger representation; also involved are sites in the left superior parietal lobe and the left intraparietal sulcus activated in number comparison—the latter site is thought to be the location of our sense of number size.[3] The authors of one recent neuroimaging study of numerical comparison and addition fact retrieval speculate that the joint activation of these (and other) areas in addition fact retrieval reflects the involvement of a finger-movement network underlying finger counting, which may, by extension, become the substrate for single-digit addition knowledge.[4] Surprisingly perhaps, the language areas of the brain, the regions around the left perisylvian fissure devoted to speech and comprehension, are not activated in addition fact retrieval. So far from being language based, the origin of our knowledge of simple sums seems to be a kind of finger expertise. Support for this comes from a study showing that performance on tests of finger knowledge and discrimination was the best predictor of arithmetical achievement in 5-to-6-year-old children.[5] Further support comes from the association between impaired ability to identify and discriminate between one's fingers (finger agnosia)[6] and impaired calculation ability (dyscalculia) commonly found in Gerstmann's syndrome.[7]

The epistemology of basic arithmetic

It is worth dwelling on the epistemic role of finger adding experience for a moment. When we follow a proof in a textbook, the sensory experience we have in reading the relevant bit of text is needed for the task, but that experience does not constitute our evidence for the conclusion of the proof; its role is instead to help convey what the argument is.[8] In contrast, our early experience with finger adding does have the role of evidence. It not only causes us to believe that $4 + 3$ is 7; it also provides our initial grounds for that belief.[9] Thus our knowledge of that addition fact and of the other single-digit sums is empirical knowledge. I am not claiming that knowledge of single-digit sums *must* be empirical. It is possible, though surely rare, that people later come to know these facts in a way that does not depend on the use of experience as evidence for them (or for anything from which they are inferred). But our initial knowledge of single-digit facts is definitely empirical. Assuming that our knowledge of single-digit facts remains empirical, our knowledge of multi-digit addition facts will also be empirical, as the latter depends epistemically on the former.

Kant famously denied that mathematical judgements are empirical. This is surprising given his argument in the following passage:

The concept of 12 is by no means already thought in merely thinking this union of 7 and 5; and I may analyse my concept of such a possible sum as long as I please, still I shall never find the 12 in it. We have to go outside these concepts and call in the aid of intuition which corresponds to one of them, our five fingers, for instance, or, as Segner does in his *Arithmetic*, five points, adding to the concept of 7, unit by unit, the five given in intuition. For starting with the number 7, and for the concept of 5 calling in the aid of the fingers of my hand as intuition, I now add one by one to the number 7 the units which I previously took together to form the number 5, and with the aid of that figure [the hand] see the number 12 come into being.[10]

His argument for denying that this is empirical knowledge ignores the details of knowledge acquisition. Rather it is a quite general argument about knowledge of necessary truths:

mathematical propositions, strictly so called, are always judgements *a priori*, not empirical; because they carry with them necessity, which cannot be derived from experience.[11]

Kant seems to have assumed that knowledge of a mathematical truth such as an addition fact and knowledge of the necessity of that truth have the very same ground. From this implicit assumption and Kant's explicit assumption that knowledge of a necessity cannot have an empirical basis, Kant's conclusion follows: knowledge of a mathematical truth will be non-empirical. But the implicit assumption (that knowledge of a mathematical truth and knowledge of its necessity must have the same ground) is very far from obvious. We may have empirical grounds for judging that $7 + 5 = 12$ and perhaps non-empirical grounds for judging that *if* $7 + 5 = 12$, it is necessary that $7 + 5 = 12$. We may then come to know that it is necessary that $7 + 5 = 12$ by deduction from those two judgements combined, and as one of them is empirical, our knowledge of the necessity of this addition fact would also be empirical. Schematically, using "□" for "it is necessary that", we may reason

$7 + 5 = 12$ (known empirically)
$7 + 5 = 12 \rightarrow \Box[7 + 5 = 12]$ (known non-empirically?)
therefore
$\Box[7 + 5 = 12]$.

Kant's argument rests on a false premiss, if knowledge of the necessity can be acquired in this way. To show that it can, we need only show how it is possible to come to know the conditional second premiss. I will turn to that task a bit later. For now, let us take the evidence at face value and accept that single-digit addition facts can be known empirically, and often are.

Unlike addition facts, we learn single-digit multiplication facts by rote. But we also come to recognize many of them to be true as repeated additions. Think for example of your knowledge that three twos make six.

Suppose, in contrast, that you only know that six sevens make forty-two from tables supplied by a teacher or some other authority. That authority may only have known it on the basis of some prior authority. But any chain of authority that yields knowledge must eventually end with some authority who knows it independently, and that authority's knowledge will be epistemically dependent on knowledge of addition facts. Given that that authority's knowledge of addition facts is empirical, so will be her knowledge of the relevant multiplication fact. Hence your knowledge of multiplication facts will also be empirical, since that knowledge depends epistemically on your own or someone else's knowledge of addition facts, and that knowledge is empirical. So our basic arithmetical knowledge is empirical. In contrast, our basic knowledge of geometry, I believe, is not empirical but conceptual (in the manner set out in Chapter 3). If this is right, the situation is the reverse of what is often believed, namely, that basic geometrical knowledge is empirical while arithmetical knowledge is a matter of pure reason.

However, our basic arithmetical knowledge is not, as Mill claimed, arrived at by inductive generalization from experience.[12] If it were, we could increase evidential confirmation by repeating the experiment making relevant variations. The point is made well by Jaegwon Kim:

Similarly, in the case of "$3 + 2 = 5$" it would be senseless to reason: "Well, now that I have checked this out with my fingers, I should also check this out with my toes, and maybe some pebbles, too, to make double sure. ..." There may be a point in repeating the experiment by counting the fingers again, but this was to ensure that the counting was done correctly, that is, to make sure that there *was* a "positive instance"; it is not to multiply positive instances.[13]

Why don't we feel that we should accumulate and vary positive instances? To put the question in another way, why do we believe that if it is true for *one* kind of thing on *one* occasion that three things and two other things add up to five things, then it is true for *every* kind of thing on *every* occasion? There is nothing special about the numbers of this example. We have the same view of all addition facts: if an equation of the form $n + m = k$ has a positive instance it cannot have any counter-instance. The reason, I suspect, is that most people have a tacit appreciation of two principles that clearly warrant this view of addition facts. Writing "|A|" for the number of members of A, and "A ∪ B" for the union of A and B, these principles can be stated thus:

For any disjoint sets S and T, and any disjoint sets X and Y

(i) $|S| + |T| = |S \cup T|$, and
(ii) if $|S| = |X|$ and $|T| = |Y|$, $|S \cup T| = |X \cup Y|$.

Together these entail that

(iii) if $|S| = |X|$ and $|T| = |Y|$, $|S| + |T| = |X| + |Y|$.

The latter guarantees that if a cardinal equation of the form $n + m = k$ has a correct instance it has no counter-instance. (For suppose that $|S| = n$ and $|T| = m$ and $|S| + |T| = k$; now let X and Y be any other sets, indeed any other *possible* sets, such that $|X| = n$ and $|Y| = m$. It follows from (iii) that $|S| + |T| = |X| + |Y|$, hence that $|X| + |Y| = k$.) In most cases this knowledge is not explicit. It arises from our concepts of cardinal number, set (or collection), and union.[14] Other principles of cardinal arithmetic can be known in the same way; in particular, the commutativity and associativity of cardinal addition flow directly from the commutativity and associativity of set union. This tacit knowledge, I claim, is not empirical but conceptual. If this is right, our knowledge of cardinal arithmetic has a mixed character: it comprises both empirical knowledge of simple equations and non-empirical knowledge of general principles.

Now let us return to the knowledge of the necessity of an arithmetical equation. We can know that $7 + 5 = 12$ in the way admirably described Kant, by means of finger counting. If we can also know the conditional

$$7 + 5 = 12 \rightarrow \Box[7 + 5 = 12]$$

we can come to know the necessity of the equation by *modus ponens*. So it remains to show how we can know the conditional. Here is how. Suppose an equation of the form $n + m = k$ is true. It will have at least one true instance—think of, e.g., the set of numerals for numbers from 1 to n inclusive, and the set of numerals for the numbers from $n + 1$ to $n + m$ inclusive. It follows from the bracketed argument given above that it will have no *possible* counter-instance, which is what it is for the equation to be necessarily true. Hence, for any simple addition fact, we can conclude that if it is true it is necessarily true. Kant's argument, then, fails. Kant was right to say that our knowledge of the truth of the equation *carries with it* the knowledge of its necessity, but this may be due to a combination of knowledge of the particular truth with *a priori* knowledge, perhaps tacit, of the general principle (iii) of cardinal arithmetic.

The decimal place system

Counting, especially finger counting, probably develops our sense of cardinal size up to 10. The special devices needed to represent numbers greater than 10 up to 20 on our fingers probably also marks out the first decade as special, although these representations are probably not much used today.[15] Eventually we get an awareness of the numbers up to 100 as divided into decades. No doubt verbal recitation of numbers beyond 20 makes a large contribution to our grasp of the decade structure of the numbers up to 100. This is probably consolidated when we learn the decimal place system of numerals. Noticing that the double-digit numerals run in sequences in which the left digit stays constant while the right one runs through the single digits is likely to aid grasp of decade structure, especially where the number word system of one's natural language does not:

Twelve   12
Quatre-vingt onze 91

Once we know the double-digit numerals we can come to recognize that a number designated by a double-digit numeral is the sum of a number of TENS and a number of ONES. This, I feel, is an underrated achievement. Understanding "32" entails understanding what three TENS is, which in turn entails understanding what kind of thing a TEN is and how it can be that there are number of them. Of course there is nothing special about TENS in this respect. In acquiring a grasp of multiplication we come to understand talk of five TWOS, three FOURS, and so on.

We go on to learn that places from right to left are for ONES, TENS, HUNDREDS, THOUSANDS. This becomes illuminating and readily generalizes when we come to know that HUNDREDS are TENS of TENS, and THOUSANDS are TENS of HUNDREDS. Without explicit instruction we pick up the general principle: if a place is for Xs, the place next to the left is for TENS of Xs. Each multi-digit numeral becomes not merely a name of a cardinal number, but also an informative description of it. For example, we can see right away that 2573 is the sum of 2 THOUSANDS, 5 HUNDREDS, 7 TENS, and 3 ONES, and that it is not a multiple of 2 or 5; further information comes with a little bit of calculation. We also have an immediate but very rough grasp of its size by its digit length. It is a four-digit number, which tells us that it is in the thousands but less than ten thousand. In other contexts we might read that a certain person has a six-figure salary or was awarded seven-figure damages. And when we have to compare numbers with numerals three or more digits long, we will probably employ an algorithm that depends on the place system: compare digit-lengths, or digits at the leftmost place at which numerals differ when digit-lengths are the same.

The importance of visuo-spatial abilities for the place system is obvious. Although the visual systems used have not been precisely mapped out, it is clear that to extract the information in a multi-digit numeral we need to recognize its constituent digits, which involves perception of shape and orientation; then the digits must be seen in line, and the place of each digit in the right-to-left ordering must be seen. So those systems responsible for perception and production of horizontal alignment and right/left order are required for facility with the place system. I mentioned earlier that dyscalculia and finger agnosia are often found together in Gerstmann's syndrome. It is possible that when this occurs finger agnosia causes dyscalculia, given the prominence of finger counting in acquiring our store of single-digit addition facts. But Gerstmann's syndrome also includes handwriting disability (dysgraphia), frequently involving an inability to write in straight lines, and sometimes also an inability to tell left from right. These impairments might obstruct the use of writing multi-digit numerals and forming visual images of them (as well as of a number line), thereby preventing mental calculation using the place system of numerals.

Multi-digit addition

Single case studies of brain-damaged patients with calculation impairments suggest that elementary calculation ability has several components over and above understanding the numerals and the operation signs for addition and multiplication. For example, Elizabeth Warrington reported the case of someone who had an intact ability to use algorithms for multi-digit calculation but had impaired single-digit arithmetic fact retrieval.[16] There are three main components:

- Storage and retrieval of number facts e.g. $3 + 4 = 7$, $2 \times 3 = 6$.
- Deployment of rules, e.g., $n + 0 = n$, $n \times 0 = 0$, put "0" on the right end of the numeral for n to get the numeral for $n \times 10$.
- Application of procedures for basic functions on multi-digit numbers using the place system.

Let us look at an instruction set for a standard addition procedure:

{1} Put the numeral for one summand beneath the other, right-aligning them, and aligning their digits in columns from the right.
{2} Check whether all columns have been summed: if NO, sum the rightmost column not yet summed, then $\Rightarrow$ {3}; if YES $\Rightarrow$ {6}.
{3} Put the rightmost digit of the sum beneath the column summed.
{4} Check whether the sum is double digit: if NO $\Rightarrow$ {2}; if YES $\Rightarrow$ {5}.
{5} Check whether there is a column of digits to the left: if YES put a "1" at top of next left column, then $\Rightarrow$ {2}; if NO put a "1" in the same row as the sum's rightmost digit and in the column just left, then $\Rightarrow$ {6}.
{6} Read the multi-digit numeral in the bottom row as the sum, then halt.

Each of these instructions, with the possible exception of {4}, involves some visuo-spatial cognitive ability. Perhaps a corresponding tactile procedure could be taught to blind children. But fully sighted children are taught an essentially visuo-spatial procedure. I stress this because it is easily assumed that the use of visual abilities in arithmetic is peripheral (as in the reading and writing of numerals), and that arithmetical thinking is essentially non-visual. Examination of the standard procedure for multi-digit addition shows that this is not always true. The standard procedures for other basic operations on multi-digit operands are also visuo-spatial. There is no way in which calculation abilities resting on standard multi-digit procedures could remain intact with a loss of relevant visual abilities. If one loses any of them one would have to learn different procedures to compensate. What are these visual abilities? I would expect at least the following to be involved:

130   VISUAL ASPECTS OF CALCULATION

- Perception of shape and orientation, to recognize and distinguish, e.g., 6 from 9;
- Perception of position in right-to-left sequence, to avoid confusing, e.g., 81 and 18;
- Perception of vertical alignment, to avoid confusing TENS with ONES;
- Perception of horizontal alignment, to avoid confusing operand and result digits.

Multiplication

Some idea of the importance of these abilities can be gleaned from the case of a patient with impaired calculation ability linked to visual deficits described in 1931. The patient had a bilateral parieto-occipital lesion. The patient had a severe drawing and pointing disability. Reading was disturbed by occasional omission of syllables or of a line; writing had intrusions of letters and irregular horizontal lines. Reading and writing two-digit or three-digit numbers was spared; but written addition of two-digit numbers was not:

The patient started with summing up the left column from top to bottom, continued with the right column in an upward direction, repeatedly lost his place, and finally gave up. In writing down the intermediate results (sums of ones and tens), he frequently confused the position values of digits. Similar failures were observed with the subtraction of 2-digit numbers.[17]

The commentator remarks that these errors are explicable by the patient's spatial deficits. But he goes on to say "mental calculation was also severely impaired" as though this was something independent of visuo-spatial ability. The reason for this comment was that when the patient was asked to multiply 12 by 13, he found the partial products ($10 \times 13 = 130$ and $2 \times 13 = 26$), but did not know whether these should be added or multiplied. In my view this impairment cannot be assumed to be independent of the patient's visuo-spatial deficits. This is because our understanding of the relationship between multiplication and addition may have a visual component. Let us look into this.

Four common ways of grasping this relation are by thinking of $k \times n$ as follows:

(a) a row of k occurrences of the numeral for n separated by "+" signs;
(b) a rectangle formed from k columns of n squares, all equal;

(c) *k* columns of *n* dots, all equal sized;
(d) *k* end-to-end *n*-unit segments along the number line.

If the patient had no visuo-spatial deficit, he could have used (a), by thinking of 12 × 13 as expressed by a string of 12 occurrences of "13" conjoined by plus signs, and could then have noticed that the partial products must be added:

$$12 \times 13 = \overbrace{13 + 13 + \ldots\ldots + 13 + 13 + 13}^{12 \text{ occurrences}} =$$

$$\underbrace{13 + 13 + \ldots\ldots + 13}_{10 \text{ occurrences}} + \underbrace{13 + 13}_{2 \text{ occurrences}} = (10 \times 13) + (2 \times 13).$$

The other visual ways of representing multiplication of two positive integers also enable one to tell that the partial products must be added.

Neuroimaging results suggest that visual areas of the brain are recruited in multi-digit multiplication, while the perisylvian language areas (Broca's and Wernicke's areas) are relatively deactivated.[18] Retrieval of single-digit multiplication facts involves regions found to be activated also in naming of tools and animals represented in pictures; and it involves a fronto-parietal network that probably underlies finger-counting representations. As one would expect, frontal networks involved in working memory and executive processes show greater activation in multi-digit multiplication than single-digit multiplication fact retrieval. Activations in multi-digit multiplication not found in retrieval include first an area within the region for working memory that is activated when maintaining visuo-spatial representations (in the left superior frontal gyrus); this area might be devoted specifically to visuo-spatial working memory, but this is quite uncertain. Secondly, bilateral areas belonging to the ventral pathway of visual processing (in the inferior temporal gyri) are activated during multi-digit calculation; these areas have been found to be activated in visual imagery of objects in the absence of any retinal input. This concurs with experimental results and introspective reports indicating that we do use visual imagery in multi-digit calculations.

Spatial format and visual imagery in calculation

The spatial format of a calculation procedure is remembered and deployed when solving problems. Evidence for this comes from interference effects when the problem is presented in a way which differs from a familiar taught format. For example, in a rare study of the use of visual imagery in simple arithmetic, evidence was found that the performance of students who had been taught a procedure for long division with the divisor placed on the left was slower when the problem format put the divisor on the right.[19] I expect similar effects could be revealed with multi-digit subtraction: if the procedure you were taught has the subtrahend below the number to be subtracted from, you will probably be slower in answering subtraction problems in which the subtrahend is placed above.

Evidence was also found indicating that internally generated visual images were often *integrated* with visual images produced by the physical inscription of the problem. A typical use would be to indicate "borrows" and "carries" in addition and subtraction problems. In a speculative conclusion, the author of that study suggests that visual imagery is important in mental arithmetic as "it acts as a surrogate for the external visual cues which the problem solver would generate if he had pencil and paper".[20]

Digital problem-solving techniques

Besides the standard multi-digit calculation procedures which exploit the place system of numerals, many other procedures taught in secondary school arithmetic[21] have a visuo-spatial component. Here are a couple of reminders. To *simplify terms* involving division, one cancels matching numerals above and below a dividing line, or above and below across a multiplication sign, by putting a slash through them:

$$\frac{\cancel{7} \times 4}{3 \times \cancel{7}} \times \frac{9}{4 \times 16} = \frac{\cancel{4}}{3} \times \frac{9}{\cancel{4} \times 16} = \frac{9}{3 \times 16}$$

Again, in *equation solving* there are several spatial-symbolic operations, e.g. moving a term across an equals sign and changing its preceding sign (+ to − or vice versa); moving a factor across an equals sign and below a dividing line:

$$3x + \widehat{15} = 45 \quad \Longrightarrow \quad \widehat{3}x = \frac{45 - 15}{}$$

Although the moves are automatic and we do not need to recall justifying principles, our cancelling practices are probably constrained by some understanding of what is going on. For example, we would not generalize our practice to include cancelling above and below across a subtraction sign as opposed to a multiplication sign. Similarly for the equation solving short cuts: we know that moving the "15" across the equals sign (and prefixing a minus sign) must precede positioning the "3" beneath the figure on the other side of the equals sign.

Analogue problem-solving shortcuts

These digital techniques are standardly taught. Schoolchildren may also pick up several analogue techniques. I will run through some now.

Clock face: Some numbers can be easily represented as a portion of a circular clock face covered by the minute hand, reckoning in minutes. For example, the value of 45 − 15 can be quickly determined using the clock face, as illustrated.

Figure 7.1

Compass: Without calculation one can find 270 ÷ 6 using one's knowledge of the compass representations of 270° and 45°.

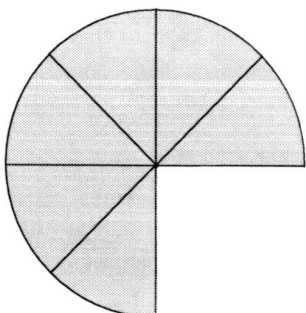

Figure 7.2

Horizon: Hayes reports an analogical use of imagery to find the sign (positive or negative) of the solution to a problem, separate from the

134   VISUAL ASPECTS OF CALCULATION

image of the problem format.[22] To solve $x + 13 = 8$, the subject calculated $13 - 8$ as 5, then visualized a vertical bar representing 8 on a horizontal zero-line, then a longer bar for 13 parallel to the 8 bar and top-aligned with it. Since the longer bar projected below the zero line, he decided that the answer must be negative.

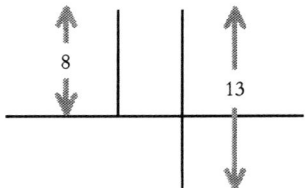

Figure 7.3

Number line: Having calculated the difference between 8 and 13, one could instead visualize a horizontal number line with a zero-point with quantities represented by vectors, eight units rightward from 0, thence thirteen units leftwards. The result is the position of the endpoint.

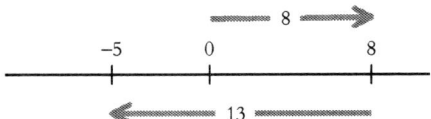

Figure 7.4

In all these cases, both the symbol manipulation techniques and the analogue short cuts, imagery could be and, I would guess, often is used in place of actual drawing and seeing.

Conclusion

The intention of this brief survey is to dispel the idea that visual thinking is peripheral in arithmetic, and to show that the normal acquisition and exercise of numerical skills and understanding makes central use of several visual abilities. In brief, visuo-spatial abilities have an important place in number perception, counting, multi-digit number representation, finger addition, algorithms for multi-digit calculation, and digital and analogue techniques for arithmetical problem solving. The concern so far has been with numerical equations rather than general theorems of number theory. The next chapter will examine the epistemology of a kind of visual thinking that seems to deliver knowledge of general theorems of number theory.

Notes

1. Fuson (1988, 1992).
2. This is adapted from Fuson and Kwon (1992).
3. Number comparison may be involved in addition fact retrieval. See Butterworth et al. (2001).
4. Pesenti et al. (2000). For background and elaboration of this hypothesis, see Butterworth (1999: ch. 5).
5. Fayol et al. (1998).
6. Kinsbourne and Warrington (1962).
7. Kinsbourne and Warrington (1963); Rourke (1993).
8. This is consistent with the possibility that we use the experience in reading the text as evidence for *something*. The point is that in normal cases we do not use the experience as evidence for the conclusion or for any premiss or implicit assumption of the argument used in reaching the conclusion.
9. Some philosophers will deny that we have such knowledge, on the grounds that we have no knowledge that there are such things as cardinal numbers. My response is given in Giaquinto (2001).
10. Kant (1781–7: B15–16).
11. Kant (1781–7: B14).
12. Mill (1843: II. vi. 2).
13. Kim (1981).
14. More precisely, we get this knowledge from activation of reliable belief-forming dispositions that come with possessing those concepts. This, I acknowledge, is a large claim, one that cannot be quickly and easily substantiated. But the kind of thing I have in mind is spelled out in Ch. 3.
15. We know that prior to the introduction of the arabic numeral place system in Europe, people used systems of hand and finger representation and calculation for numbers well into the thousands. See the chapter 'Finger Counting' in Menninger (1969).
16. Warrington (1982). For a review of clinical studies of dyscalculia, see Cipolotti and van Harskamp (2001).

17. Ehrenwald (1931), cited in Hartje (1987).
18. Zago et al. (2001). I do not know of other neuroimaging data that bear on this question.
19. Hayes (1973).
20. Ibid.
21. Secondary school in the UK and high school in the USA are roughly the same.
22. Hayes (1973).

8
General Theorems From Specific Images

Our visual abilities can help us solve numerical problems in a variety of ways. This was highlighted in the previous chapter. But these were problems with a specific numerical answer. Means that are extremely useful for finding the numerical value of an arithmetically composite term may be pretty useless for discovering a general formula. Compare, for instance, the tasks:

(a) Find the number of 2-element subsets of a 4-element set.
(b) Find the formula for the number of n-element subsets of a k-element set, when $1 \leq n \leq k$.

To solve task (a) one can make or visualize 4 dots; then draw or imagine lines pairing them off in all possible ways, keeping count of the acts of actual or imagined line-construction. Nothing like this will work for task (b). Comparison with geometry may reinforce the feeling that visualizing is not a means of discovering general theorems of arithmetic. For one may come to believe a general theorem about squares by visualizing operations on a single square, and this could be reliable because all squares are the same in shape and so share all their intrinsic geometric properties. But the same line of thought does not carry over to arithmetic. No two numbers share all their intrinsic numerical properties. So if there are many numbers of a given kind, such as prime numbers or fourth powers, one should be wary of reaching a conclusion about all numbers of that kind by visualizing operations on just *one* number of things. Nonetheless I will try to show in this chapter that some general arithmetical theorems almost certainly can be discovered by visualizing.

The plan is as follows. First I describe two examples of what look like acceptable routes to general theorems in which visualization has

138 GENERAL THEOREMS FROM SPECIFIC IMAGES

a non-redundant role. Then I canvass and assess in turn two natural objections to the acceptability of these examples. Finally, focusing on a modified version of the first example, I give a positive account of how it is possible to use visual imagery together with arithmetical concepts to reach a general theorem in a reliable way.

Triangular numbers

A *triangular number* is the sum of the first k positive integers, for some positive integer k. The first triangular number is 1, the second is $1 + 2$, and in general the kth is $1 + 2 + \ldots + (k-1) + k$. Suppose you want to find the kth triangular number, for a certain k. If k is very small one can just add. Without a calculator this would be tedious if k were 20, difficult if k were 200, and a fool's errand if k were 2000. For large numbers one wants a general formula. Here is a way that one could find one.

Think of each number n as represented by a column of n dots; then visualize these representatives starting with just 1 dot on the left, then adding columns, n dots in the nth column going from left to right, with a horizontal base line—stopping when you have an array of manageable size. Let z be the number of columns. Then the zth triangular number—call it $T(z)$—is the total number of dots in this array. The task is to find a formula for the simple calculation of $T(z)$. The array described should seem to form a triangle. Figure 8.1 illustrates.

Figure 8.1

Now visualize a copy of this triangular array just to the left of the original. Then invert the copy and place it above the original, so that the z-dot column of the copy is vertically in line with the 1-dot column of the original, and the 1-dot column of the copy is vertically in line with the z-dot column of the original. Then vertically close the two triangular arrays until there is row alignment, keeping the z-dot column of one triangle vertically in line with the 1-dot column of the other. See Figure 8.2(a)–(d).

GENERAL THEOREMS FROM SPECIFIC IMAGES 139

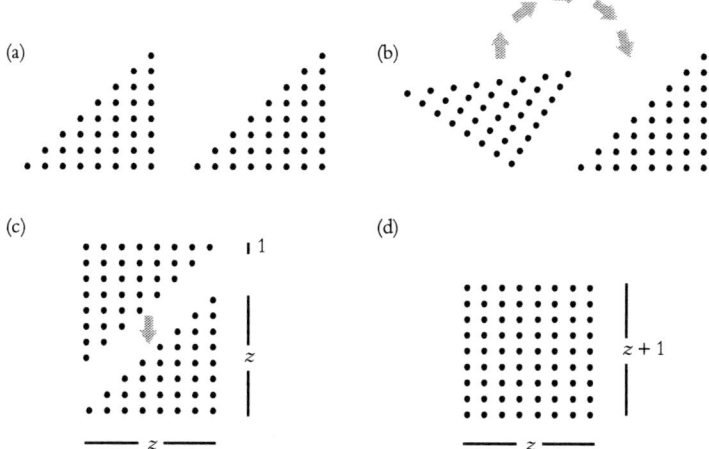

Figure 8.2

The result is a rectangular array, z columns of $z+1$ dots, containing twice the number of dots in the original triangle: $2 \times T(z) = z \times (z+1)$. Hence the general formula for the sum y of the first z positive integers is:

$$T(z) = \frac{z \times (z+1)}{2}$$

Shortly, I will examine the claim that this route to the general formula is epistemically acceptable. But first, here is another candidate.

Square numbers

A *square number* is the number $k \times k$, for some positive integer k. If we let a row of k dots represent the number k, a square array consisting of k vertically aligned rows of k dots represents $k \times k$. Using these conventions it is not difficult to see that every square number greater than 1 has one of two forms. Here is how. An even number has the form $2m$, which can be represented by a pair of horizontally aligned rows of m dots, the rows separated by a gap noticeably larger than the inter-dot gaps within rows, not too large but large enough to maintain the appearance of two separate rows, as in Figure 8.3.

· · · · · ·   · · · · · ·

Figure 8.3

Then $2m$ such rows, vertically aligned and with a gap to remind us that the number of rows is even, forms a square array of $2m \times 2m$ dots which represents that number. But visualizing or seeing this array reveals that it forms 4 square arrays, each consisting of $m \times m$ dots, as illustrated in Figure 8.4.

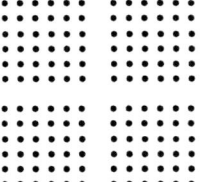

Figure 8.4

An odd number $2m + 1$ can be represented by a pair of horizontally aligned m-dot rows separated by a largish gap containing an isolated dot. $2m + 1$ of these $(2m + 1)$-dot rows, vertically aligned, with its central row isolated, forms a square array of $(2m + 1) \times (2m + 1)$ dots. But visualizing or seeing this array reveals that it forms 4 square arrays, each of $m \times m$ dots, plus 4 lines of m dots, plus 1 central dot, as in Figure 8.5.

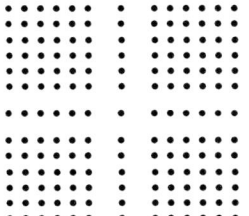

Figure 8.5

Since every square number is the square of an even or odd number, every square number greater than 1 is representable by one of these two forms. Squares of the first form are even, being composed of a quartet of equal squares of $m \times m$ dots; squares of the second form are odd, since the central dot remains when the $(m \times m)$-dot squares are paired off and the m-dot lines are paired off. Hence every *even* square number is the sum of 4 square numbers.

This apparently insignificant general theorem has an important consequence, namely, that the positive square root of 2 is an irrational number. Suppose some square number is *twice* another square number. Then there would have to be a *smallest* square number that is twice another square number, as there is no unending sequence of ever smaller square numbers. Let k^2 be the smallest: $k^2 = 2(h^2)$. Then k^2 is an even square; so, by the theorem, there is some number n such that $k^2 = 4(n^2)$. From these two equations it is obvious that $h^2 = 2(n^2)$. As h^2 is smaller than k^2, it follows that a smaller square number than k^2 is twice another square number, contradicting the definition of k^2. This reduces to absurdity the supposition that there is a square number that is twice another. The fact that no square number is twice another is easily seen to be equivalent to the proposition that the positive square root of 2 is irrational.[1]

Those are the examples to be discussed in the rest of this chapter. They are representative of a kind of *pebble* arithmetic known to be practised in ancient times by the Pythagoreans.[2] Now I will set out a couple of objections to the claim that these examples constitute means of discovery, and then I will give responses to the objections. The first of these I call the particularity objection, and the second I refer to as the problem of unintended exclusions.

The particularity objection

The arithmetical theorems put forward as discoverable by a combination of reason and imagination are *general* theorems: (i) For *every* positive integer n, the sum of the first n positive integers is a half of $n \times (n+1)$; (ii) For *every* positive integer n, if n^2 is even it is the sum of 4 square integers. But there is a natural objection to the idea that visualizing and reasoning as described can be a way of discovering a general arithmetical truth: if we succeed in visualizing a triangular array of dots or a square array, there must be a particular number of dots in the visualized array; so the most that we can discover by visualizing in the way suggested are particular instances. If, for example, we visualize an array of exactly seven columns of dots, n dots in the nth column, and proceed as suggested in the section on triangular numbers, the most that we are justified in concluding is that half of $7 \times (7+1)$ is the sum of the first seven positive integers. This is the particularity objection.

Response to the particularity objection

This objection assumes that in visualizing an array of distinct objects there must be a specific number of objects visualized; and from this assumption the objection steps to the conclusion that the only arithmetical discoveries one can make by visualizing a plurality of objects are truths about the precise number of objects visualized. I reject both the assumption and the step. First I will consider the assumption.

Imagine a room full of people, all seated facing you, as you begin to address them. Must there be a particular number n such that you have imagined precisely n faces before you? Imagine the sky on a clear night. Must there be a precise number of stars visualized? Introspection does not suggest a positive answer. So why would one think that there is a particular number of things visualized? An image of a starry sky could be an instance of a kind of texture representation, rather than a phenomenal aggregation of star images.

There are two views that could lead one to think that a particular number of objects (faces, stars, or whatever) must be represented in an image of an array of things; one of these views is primitive, while the other is more sophisticated. According to the primitive view, in visualizing something, as in seeing something, there must exist an entity that one is visualizing, and visualizing is a relation between the visualizer and the visualized object. No one thinks that when one visualizes, say, an apple, there must be some particular apple that one is visualizing; on the primitive view the object of the visualizing is internal to the mind, in this case not an apple but a visual representation of an apple. But the internal object of visualizing must, on the primitive view, share with things of the visualized kind visible properties that give those things their visual appearance, e.g., shape, colour, size, and shine. When what is visualized is a collection, say, a bunch of bananas in view, number will be one of the relevant properties: the internal object will be composed of a definite number of single banana representations. Numerical particularity falls out of this.

The primitive view does not stand up to scrutiny. An apple has roughly spherical shape, and that is something that contributes to its appearance as an apple. But is it plausible that a visual representation of an apple, which is something internal to the mind, is roughly spherical? To have an image of something as spherical and shiny, is not to have a spherical and shiny image. There is something common to seeing an X and visualizing an X, but it does not consist in a set of properties common to a seen X and a visual image of an X. Rather it consists in a set of properties common to the representational content of experiences of seeing an X and of visualizing an X.

Let us turn to the more sophisticated reason for thinking that in visualizing a plurality of things there must be some particular number such that one is visualizing just that many things. The more sophisticated reason, which may be held quite independently of the primitive view of visualizing, is as follows. In a verbal description of a scene, we can convey that some visible condition is fulfilled without conveying the specific way in which the condition is fulfilled. Joseph may have a many-coloured coat, according to the story, without any colours being specified. In contrast, it may seem plausible that a picture which conveys the information that a coat is many coloured cannot do so without also specifying some particular colours. The thought is that a coat can be depicted as many coloured only if for some particular colours, the coat is depicted as having those colours. Similarly with respect to numerical conditions: Jesus is said to have disputed with learned men, no particular number of learned men being specified; but a picture portraying Jesus disputing with learned men must show a particular number of disputants; for some particular number there must be just that number of disputants in view. These considerations seem pertinent because visual experience[3] is pictorial rather than verbal. Whether by vision or visualization, the content of the experience is conveyed in a way that a picture conveys content, not in the way that verbal narrative conveys content. So visual experience of an array of distinct dots must, it seems, represent a specific number of dots.

There are several things wrong with this line of thought, some of them important for understanding visualization. First, visual experiences, or those parts of them we call images, are not pictures, and so we cannot assume that what is necessarily true of pictures is true of all visual experiences. The analogy between visual images and representational paintings can be treacherous as well as useful. For example, there is sense to talk of *scanning* the content of one's visualizing experience or *zooming-in* on some part of it.[4] But we must be careful here, as these expressions are metaphorical. What makes these metaphors appropriate is that (a) they denote changes (of visualizing experience) which result from imagining scanning or zooming-in on something external and (b) the experience of visually imagining scanning or zooming-in on something external is subjectively similar to the experience of actually scanning or zooming-in on something external. But one way in which the analogy breaks down is that one can zoom-in on a part of a picture without changing the picture, whereas one cannot zoom-in on part of the content of a visualizing experience without changing the experience. If the number of dots in a visualized array is not apparent to you the visualizer, there is no mental act you can perform, such as "zooming-in", "scanning" from left to right and counting as you go, by which you can determine the number of dots in the array originally visualized, because the original experience is not preserved in the process. The central point, here, is that even if it were true that a *picture* never represents a plurality unless there is a particular number of things it represents, a *visual image* might be pictorial in ways which do not entail that kind of numerical determinacy. They might, for example, be pictorial in the following two ways:

(1) Every representing part of an image of an object O represents a part of O;
(2) Order relations on an imaged surface of O are preserved by corresponding relations among representing image parts.[5]

These do not entail numerical determinacy of images.

A second problem for the argument for numerical determinacy in images is that pictures can depict a plurality of things of a certain kind without there being a particular number n such that the picture depicts exactly n things of that kind. In Frans Hals's jolly portrait *Yonker Ramp and His Girlfriend*, the left sleeve of the young woman is shown with a line of buttons. Exactly how many buttons are in view? It seems to me from looking at the painting that there is no answer to this question. Similarly for Yonker's exposed teeth; a plurality but no particular number of teeth is shown. The same point holds for colours or at least shades of colours.[6] A painting can represent a surface in dim and unusual light; although it represents the surface as having some shade there need be no particular shade that the surface is represented as having. So it is in general mistaken to think that whenever a picture represents some individual or collective item as having some property in a property spectrum, such as colours or cardinal numbers, there is a particular property in the spectrum that the item is depicted as having.[7]

For these reasons it seems to me that the numerical determinacy assumption is untenable. But even if it were correct, the conclusion drawn in the particularity objection does not follow. For we might reach a general truth from a particular example by abstracting certain features and ignoring some details. The abstracted features might be sufficient as ingredients for a reliable procedure for reaching a true general belief, if all members of the domain of generality shared the abstracted features. The point in this case is that even if the visualized array were numerically determinate, the particular number of dots in the array might not be used in the procedure, while those features that *are* used might be common to all possible displays of the relevant type, (e.g., those conforming to the general instruction to visualize an array of dots composed of a number of columns, n dots in the nth column going from left to right, with a horizontal base line). So the particularity objection fails.

The problem of unintended exclusions

The second objection stems from a problem already encountered in Chapters 4 and 5, the problem of unintended exclusions.[8] Recall the example from those chapters. This is a way of reaching the belief that the area of a triangle with a horizontal base is half its height times its base. Imagine a triangle ABC with a horizontal base in a rectangle on the same base and with the same height. Drop a perpendicular BF from the upper vertex of the triangle, to see that the triangle is composed of parts ABF and CBF, each being half of rectangles $ADBF$ and $CEBF$ respectively, which together compose the entire rectangle $DECA$: So $ABC = ABF + CBF = ADBF/2 + CEBF/2 = DECA/2$. See Figure 8.6 (left). This does not work for triangles with an obtuse angle at the base. See Figure 8.6 (right).

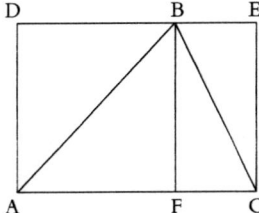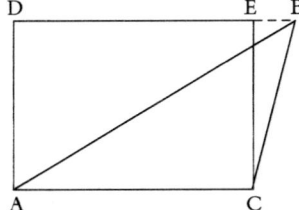

Figure 8.6

In using this method we have unintentionally excluded triangles that have an obtuse angle at the base. For this reason, the true general belief that the area of a triangle with a horizontal base is half its height times its base cannot be discovered in this way. The unintended exclusion vitiates this way of arriving at the belief.

Do the visual ways of reaching the theorems about triangular and square numbers have unintended exclusions? Yes. The exclusions will differ depending on whether the images of arrays of dots are or are not numerically determinate. If the images one forms are numerically determinate, large numbers will be excluded. People cannot simply see, when focusing on a 47-dot array, that it has exactly 47 dots. This inability need not be due to blurring or proximity of neighbouring dots; it persists when the dots are sharp, when each dot is clearly distinct from the background and from its neighbours. This fact has a correlate in our powers of visual imagination: we are typically unable to carry out the instruction to visualize a triangular array of exactly 47 dots. The best we can do is to visualize an array of *roughly* 47 dots and the result will be subjectively indistinguishable from visualization of roughly 46 dots. The same applies to all larger numbers. So numbers this large and all larger numbers are excluded when the images are numerically determinate. If the images are *not* numerically determinate small numbers are excluded. We are typically unable to visualize an array of roughly but not exactly 3 distinct dots. If we visualize an array of dots of no specific number, we would not describe what was visualized by saying that it *might* be a 3-dot array, but it might have slightly more or slightly fewer than 3 dots. This is because the appearance of an array of 3 dots or fewer is typically incompatible with numerical indefiniteness. When the dots form a pattern, such as a square, this probably extends to 4 dots. Definite or indefinite, then, visual imagery of the kind described earlier will result in unintended exclusions, and this stops the visual route to belief being a means of discovery. This is the second objection.

Three grades of unintended exclusion

My response to this is to distinguish between the following epistemic situations. First is the situation in which the procedure followed in visual imagination can still be applied to the excluded cases. Exclusions of this sort are *epistemically harmless*, for if these are the only exclusions the procedure may be reliable and no violation of epistemic rationality need be committed

GENERAL THEOREMS FROM SPECIFIC IMAGES 147

in reaching the general belief. Suppose that we used the method mentioned above to reach the belief that the area of any *acute-angled* triangle with a horizontal base is half its height times its base. Then an example of a harmless exclusion is that of triangles with a base angle only just less than a right angle (illustrated in Figure 8.7).

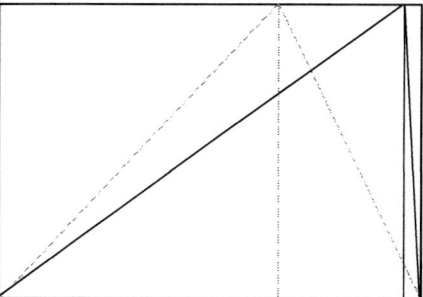

Figure 8.7

In the second kind of situation the procedure cannot be applied to the excluded cases, but there is some alternative way of dealing with them, so that the procedure supplemented by the alternative for the excluded cases is an epistemically acceptable way of reaching the general theorem. In this circumstance exclusions are *vicious but corrigible*. An example is the exclusion of triangles with an obtuse angle at the base mentioned earlier. Here is a sketch of an alternative. Set the obtuse angled triangle in a parallelogram whose diagonal is the triangle's longest side, and which shares two sides with the triangle, as illustrated in Figure 8.8 (i).

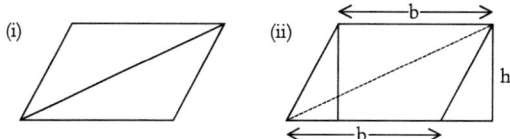

Figure 8.8

The area of the triangle is half of the area of the parallelogram, as the latter is composed of two congruent parts, one of which is the original triangle. Now by dropping perpendiculars to the base line from the upper vertices of the parallelogram, two right-angled triangles are formed, as illustrated in Figure 8.8 (ii). As the two right-angled triangles are congruent, the area of the parallelogram equals the area of the rectangle with horizontal length b and height h; so the area of the parallelogram is h × b. But h is the perpendicular height of the original triangle and b is the length of its base. Hence the area of the triangle is half its height times its base.

So the first two kinds of exclusions are harmless exclusions and corrigible exclusions. Exclusions of the third kind are *incorrigibly vicious*. To illustrate, we can use a fallacious argument already paraded in Chapter 5. This is the inductive argument for the claim that no two horses differ in colour. The error there is a manner of thinking that cannot apply to sets of just two horses (i.e. the case when $2 = n + 1$ in the induction step). As there is no way of supplementing the argument for this case, the exclusion is incorrigible.

To assess the damage done by unintended exclusions in the examples concerning triangular and square numbers, we need to find out which grades of unintended exclusion are involved: harmless, vicious but corrigible, or incorrigibly vicious.

Unintended exclusions of triangular and square numbers

Let us look first at the exclusions resulting from the visual route to the formula for triangular numbers. The exclusions in this case depend on whether in visualizing an array of dots there is a determinate number of dots visualized. As mentioned before, for normal adults indeterminacy excludes the 1-dot array and the 3-dot array, the first two triangular numbers, and possibly the next one up, the 6-dot array. Determinacy excludes all large arrays—I would guess that very few of us can visualize exactly 15 dots.[9] A determinate array consisting of a single dot is also excluded, because a 1-dot array does not form a triangle of dots, and so the procedure cannot be carried out.

To see which if any of these is harmful, it will help to step back a moment to look at how visualization is being used. There are two possibilities. One is that visualizing is used to provide observational data to serve as evidence for the belief arrived at. The second is that it is used to bring to mind a type of procedure applicable to any array of a certain kind. It is the second use of visualizing that I am proposing as part of a means to discovery. The procedure brought to mind is applicable to any collection of dots arranged in the manner described, that is, as columns of dots, n dots in the nth column ascending from left to right, with equal spacing between and within columns, so as to form a triangular array. What the visualizing reveals is that *if* a number, any number, of dots can be arranged to form such a triangular array, twice that number of dots can be arranged to form a certain rectangular array. From this and prior beliefs we extract the formula. So it does not matter that some triangular number is excluded by what is visualized, as long as that number of dots can be arranged to form a triangular array as described above. As 3 dots can be so arranged, the exclusion of a 3-dot array is harmless: the visualized procedure still applies. The same goes for any larger triangular numbers that are unintentionally excluded.

But 1 dot cannot be arranged to form a triangular array.[10] So the exclusion of 1 dot is vicious and the suggested way of arriving at the formula does not ensure that the formula works right at the beginning. But of course the exclusion is corrigible: the sum of the positive integers up to and including 1 *is* equal to a half of $n \times (n+1)$ when $n = 1$. This we know by substituting "1" for "n" in the expression and calculating the result. It is not revealed to us by visualizing a triangle of dots, etc. Hence this way of visualizing is not a way of discovering that for *every* positive integer n, the sum of the first n positive integers is a half of $n \times (n+1)$; but it might be a way of discovering this for all n greater than 1. This is a general theorem, even if not the completely general theorem we originally had in mind; so the problem of unintended exclusions does not bar discovery of general arithmetical truths by visualizing in every case. For the completely general theorem we must make special provision to check that the formula works in the initial case.

What about unintended exclusions of square numbers? Recall the combination of visualizing and reasoning presented earlier by which we can reach the theorem that for every positive integer n, if n^2 is even it is the sum of 4 square integers. In the process, visualizing leads one to the theorem that for every positive integer n greater than 1 there is a positive integer m such that n^2 equals either $4(m^2)$ or $4(m^2) + 4m + 1$.

Whether or not there is a determinate number of dots in the array one visualizes, the procedure excludes the cases when n^2 is 4 or 9. Let me explain why. The smallest square number we have to consider is 4. Visualizing in this case is supposed to reveal that if any number of dots greater than 1 can be arranged to form a square, it can be arranged to form a square consisting of *either* 4 equal squares of dots *or* 4 equal squares of dots plus 4 equal lines of dots plus a single dot. But the visualized operations do not reveal this. Being even, a square of 4 dots would have to consist of 4 equal squares of dots. But how many dots would be in each of these 4 subsquares? The number could only be 1, and 1 dot cannot be arranged to form a square of dots. So 4 is a counter-example to the proposition about form that was supposed to have been revealed by visualizing. The same is true of 9 and for the same reason. Being odd, a square of 9 dots would have to consist of 4 squares of m^2 dots plus 4 lines of m dots plus 1 central dot. In this case m can only be 1; hence m^2 can only be 1. So 1 dot would have to form a square of dots, which it cannot do. (It would also have to form a line of dots, which again it cannot do.) So for a reason having nothing to do with image indeterminacy, 4 and 9 are harmful exclusions.

But there is no harm in exclusions of greater square numbers. The visualizing reveals that *if* a plural number n of dots can be arranged to form two equal lines of dots or two equal lines of dots plus a single dot (where a line of dots can consist of just two dots but not just a single dot), then n^2 dots can be arranged to form *either* 4 equal squares of dots *or* 4 equal squares of dots plus 4 equal lines of dots plus a single dot. The antecedent condition here is not met when n is 2 or 3; it is met when n is any larger number. So it does not matter if larger numbers and their squares are excluded by what is visualized, provided that those numbers can be arranged to form two equal lines of dots or two equal lines of dots plus a single dot. As this proviso is met for all larger numbers, their exclusion is harmless.

So, once we take notice of the harmful exclusions (when n is 2 or 3), the visualization can be a non-redundant part of a means of discovering the restricted theorem that every even square number greater than 4 is the sum of 4 square numbers. Though not the theorem we originally had in mind, this is a general theorem covering an infinity of cases.[11]

The positive account: images and generality

Once unintended exclusions have been considered we seem to have reliable ways of reaching general theorems with an infinity of instances. But how is this possible? In the next two sections I will try to answer this question, restricting discussion to the example about triangular numbers. Let there be no pretence that the account I shall give describes what actually happens when one arrives at the relevant belief in the manner described earlier. For that new empirical work is needed. The aim is merely to give an account of a possibility that is consistent with prevalent views in cognitive science.

In this section I will try to explain how it is possible to achieve generality when thinking with particular images. In the section after I will try to show how imagination and inference can link up to deliver the required result. First, I will say something about the nature and formation of the relevant imagery within the framework of Kosslyn's later theory.[12] This is to be regarded as nothing more than a sketch of a possibility, which I will use to give a positive answer to the question of generality.

It will be helpful to recall a distinction among uses of the word "image" mentioned in Chapter 6. On one use, an image is the subjectively accessible content of a visualizing experience, or an item in such content—a *phenomenal* image, as I call it. On another use, an image is an entity of a kind posited in theories of visual information processing to explain such findings as the rotation data of Shepard and Metzler (1971). This I call a *psychological* image, or an image, without qualification, when it is clear that I am talking about a theoretical posit. In Kosslyn's theory the visual imagery system is part of the visual system, and an image is a pattern of activity in the visual buffer.[13] In seeing, signals from retinal stimulation become input to the visual buffer, whose output becomes input to the category pattern[14] activation system, whose output becomes input to associative memory. That is "bottom-up" processing involved in recognizing an object as an apple, say, or a cup. But feedback loops in the system operate to aid visual recognition. So there are processing paths from the systems more remote (in bottom-up processing order) from the retina back down to the visual buffer. Processing in that direction is "top-down". In Figure 8.9, part of

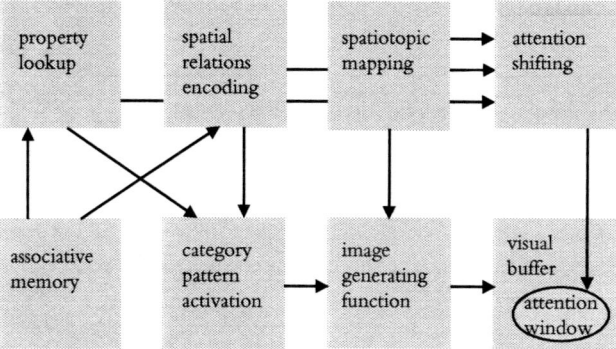

Figure 8.9. *Source*: Adapted from Kosslyn (1994)

the visual system of Kosslyn's theory is illustrated with arrows indicating only top-down processing paths.[15] Top-down processing is initiated by input to associative memory (and, when there is movement, input to the spatiotopic mapping system). The top-down processing involved in seeing is initiated by inputs that derive from retinal stimulation via the visual buffer. Visual imagery is formed when top-down processing is initiated by inputs from outside the visual system by, for example, an internal instruction to imagine an apple rolling across a table.

An internal instruction to imaging something simple may activate a representation in associative memory. Output from associative memory is processed by systems that encode property information such as colour, texture, and shape (property lookup) and spatial information such as alignment of features and locations of subparts (spatial relations encoding). This encoded information is sent to the category pattern activation system, which selects a code for a pattern that "fits best" with the information supplied. If the selected pattern code is activated strongly enough, that will stimulate a pattern of activity in the visual buffer, i.e. cause a visual image.

For complex imagery the story is more elaborate.[16] Complex imaged objects often have hierarchies of parts, such as fingernail, finger, hand, lower arm, and so on. Visualizing parts that are details of the whole requires images of greater resolution than does visualizing the whole. Typically a global image is formed first, e.g. of a whole human body, which will include some parts. The relative size and location of a part will be encoded by the spatial relations encoding system. To focus on a part, size and location information is sent to the attention shifting system, which responds by narrowing and shifting the attention window to appropriate scope and location. The attention window surrounds just that part of the pattern in the visual buffer representing the part to be focused on. As there is a scope-resolution trade-off,[17] narrowing the attention window increases resolution, which thereby achieves the objective.[18]

Some complex imagery is not formed in the global-to-local manner just described. An alternative is sequential "attention-based" image formation. A simple image may be formed at a given location; then spatial location information goes to the attention-shifting system which moves the attention window in successive steps, allowing a complex image to be formed from simple images at certain locations. Something like this *may* happen when one first tries to visualize columns of dots, n dots in the nth column, ascending from left to right. One starts with an image of a single dot; then adds an image of a "colon" of dots to the right. Then one continues adding column images in turn, each time "placing to the right"[19] an exact copy of the most recently added column image plus a dot image just above the copy; these operations are performed so as to preserve intercolumn spacing and alignment of column tops. With each step the attention window is widened so that the original configuration of dots becomes a less distinct part of the new lower resolution image. The process would stop after a number of columns, as the whole image has to be contained within the limits of the visual buffer without a decrease in resolution so great that column structure is obliterated.[20]

It should not be assumed that the phenomenology of the experience is totally determined by activity in the visual buffer. Stored spatial information also contributes, such as information that dots are equal in size, and that between-column spacing and within-column spacing is the same throughout. This information is crucial for a feature of the phenomenal image, namely that the columns of dots appear to form a right-angled triangle with a horizontal base and a vertical side.

What about numerical determinacy? The construction of the psychological image sketched here entails that it is composed of a determinate number of dot images. A non-compositional alternative is that a column or triangle or other array of dots is represented as a region of a certain shape with a certain visual texture, like a wallpaper pattern. In this case the image will not represent a determinate number of dots, and the visualizer will not be able to determine a number from the phenomenal image. If the psychological image is composed of a determinate number of dots, it is *still* possible that the number will be subjectively unavailable, hence that the phenomenal image does not represent a determinate number of dots. So this account of image construction leaves open the question of numerical determinacy for phenomenal images. But it does not matter for my purposes, as my account of how a specific image can have general significance will not depend on whether there is numerical determinateness or not.

Generality

Let us take a slightly oblique approach to the question how a specific image can bear general significance. A standard arithmetical argument for a general conclusion has interpreted formulas containing symbols called "variables", and generality is introduced via conventions governing the variables. Often the argument proceeds by transformations of expressions with variables. Here, for illustration, is an argument to show that $(a/b)^2 = a^2/b^2$.

$$\left(\frac{a}{b}\right)^2 = \frac{a}{b} \times \frac{a}{b} = \left(a \times \frac{1}{b}\right) \times \left(a \times \frac{1}{b}\right) = (a \times a) \times \left(\frac{1}{b} \times \frac{1}{b}\right)$$

$$= a^2 \times \left(\frac{1}{b \times b}\right) = a^2 \times \left(\frac{1}{b^2}\right) = \frac{a^2}{b^2}.$$

How does an argument containing variables such as this enable us to discover a general theorem? It provides a valid schema for proving all instances. We recognize that uniform substitution of numerals other than "0" for the variables would result in a valid argument whose conclusion is an instance of the general proposition. The intended scope of generality in this case is the class of positive integers. The central point is that for any chosen pair of members of the intended class, without exception, substituting their names systematically for the variables results in a proof.

When we visualize a right-angled triangular array of columns of dots, n dots in the nth column, ascending from left to right, the phenomenal image is determinate with respect to certain aspects of spatial form. What takes the place of the variable in the schema is the spatial form. The visual procedure applied to any number-determinate array of that form (plus the final arithmetical reasoning) is a reliable way of reaching an instance of the general conclusion: for any integer z greater than 1, the sum of integers from 1 to z is half of $z \times (z+1)$. Generality is gained by recognizing the spatial form as a form which a number of dots can be arranged to compose, whenever that number is a triangular number greater than 1.

How do we recognize this? Here is a possibility consonant with the speculative account of the psychological image construction given earlier. This has two parts. In the early stages we form a 3-dot image with base dots horizontally aligned and right dots vertically aligned. By the pathway from visual buffer via category pattern activation system to associative memory, this activates a representation of Right-Angled Triangle, so that we have a phenomenal image of a right-angled triangle of dots. Now going beyond Kosslyn's model, I suggest that representations in associative memory are concepts. So what is activated in this case is a perceptual concept for right-angled triangles. Thus the phenomenal image carries the information

GENERAL THEOREMS FROM SPECIFIC IMAGES 155

that the dots form a right-angled triangle. In this way we recognize that a plurality of dots whose number is the first triangular number after 1 can be arranged to compose a right-angled triangle of columns of dots, n dots in the nth column, ascending from left to right, i.e. the second triangular number of dots can have the specified form.

The next step is to recognize that if one triangular number of dots can have the specified form so can the next. This is relatively straightforward. Visualize a triangle of the specified form composed of columns of dots; then visualize adding an exact copy of the rightmost column to the right, at the distance which preserves intercolumn spacing, with a dot added to the top at a distance which preserves alignment with the tops of the columns to the left, as in Figure 8.10.

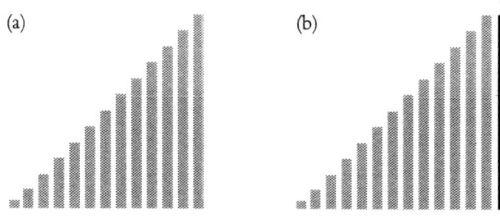

Figure 8.10

Let "$T(k)$" abbreviate "the kth triangular number". As the starting array, diagrammatically presented in Figure 8.10(a), has the specified form, it falls under the following description: for some k greater than 1, $T(k)$ dots in k columns, n dots in the nth column ascending from left to right, with a rightmost column of k dots. So by the way in which the starting array is extended (see Fig.8.10(b)), the resulting array falls under the description: $T(k+1)$ dots in $k+1$ columns, n dots in the nth column ascending from left to right, with a rightmost column of $k+1$ dots. This is apparent to the visualizer from the instruction for extending the phenomenal image: add a copy of the rightmost column i.e. a k dot column, plus a dot on top, thereby forming a $k+1$ dot column, with appropriate alignments. Given that the visualizer has an at least tacit disposition to infer according to mathematical induction, she will then be disposed to infer that for any triangular number greater than 1, that number of dots can be arranged to compose an array of the specified form.[21]

156 GENERAL THEOREMS FROM SPECIFIC IMAGES

The positive account: imagery, inference, and discovery

As with the geometrical examples discussed in earlier chapters, the role of visualizing is not to provide sensory evidence, but to activate certain inferential dispositions involved in possessing a concept. One might say that the role of visualizing is conceptual rather than empirical. The pivotal concept in this case is a concept of finitely iterated cardinal addition, {**sum**}, characterized by the following inference forms:

$S_1, \ldots, S_z$ are disjoint sets of cardinal numbers $k1, \ldots, kz$ respectively

---

The cardinal number of the union of the S_j, $1 \leq j \leq z$, is **sum** $(k1, \ldots, kz)$

The cardinal number of T is **sum** $(k1, \ldots, kz)$

T is the union of disjoint sets of cardinal numbers $k1, \ldots, kz$ respectively.

Explanatory details about this are given in the endnotes.[22] Since disjoint columns of dots are disjoint sets and the number of dots in the nth column from the left is n, application of the first inference form yields the conclusion that the number of dots in the z columns combined is **sum**$(1, \ldots, z)$, the sum of the first z positive integers. Thus the visualizer can take the triangular form to represent an array of $T(z)$ elements, i.e. **sum**$(1, \ldots, z)$. The experience of visualizing is not used as evidence as to how things are or could be in the physical world. Nor is image inspection needed to determine the applicability of the concept {**sum**} here, since this is supplied by the content of the phenomenal image, which is partly determined by information involved in generating the psychological image.

The next stage in the process is the generation in visual imagination of a copy of the triangular form so that two such forms juxtaposed in the same plane are represented.[23] The copy is then rotated and placed above the original so as to get the rectangular form of columns of dots, each column having the same number of dots. There are two motion-added transformations[24] here, rotation and relocation, achieved by altering instructions to the spatio-topic mapping system.[25] This combined transformation occurs so that represented spatial properties of the copy are preserved.

What could motivate this crucial transformation? Prior knowledge will be activated by attempting to solve the problem. This could include the

GENERAL THEOREMS FROM SPECIFIC IMAGES 157

fact that the number of items in a collection is easy to compute if the items form a rectangular array with a known number of columns and a known number of items in each column. An occurrence of this thought might then produce an attempt to "complete the rectangle" of which the original triangle is a half. Once this is done the number is computed. The concept used for this is a concept {×} for binary multiplication, individuated by the following inference form and its converse:

S is the union of n disjoint sets each of cardinal number k
___
The cardinal number of S is $n \times k$.

In order to apply this, one must draw on the precise placing of the copy over the original triangular form: the longest column of the copy over the smallest of the original and the smallest of the copy over the longest of the original, i.e. a z-dot column over 1 dot on the left and 1 dot over a z-dot column on the right, so that the leftmost (and rightmost) column of the rectangle has $z + 1$ dots. Given this, and given that each column in the rectangle has the same number of dots and that there are z columns in the rectangle, one can infer that the rectangle is the union of z disjoint columns of dots each of cardinal number $z + 1$. Then inferring in accord with the inference pattern for {×} given above, one concludes that the number of dots in the rectangular array is $z \times (z+1)$.

There is a danger here. If one visualizes the operation on triangles with continuous edges and sharp vertices, one might omit to add the 1 dot of the shortest column and so seem to end up with a rectangle of $z \times z$ dots. This can be avoided by maintaining images of discontinuous triangular arrays with single dots constituting the "vertices". Alternatively, one can keep alive the instruction to place the z-dot column of the copy over the 1-dot column of the original (or the 1-dot column of the copy over the z-dot column of the original) when calculating the number of dots in left (or right) column of the newly formed rectangle of dots.

Discovery comes by viewing the final form in two ways, both as a rectangle of equal columns and as composed of the two equal triangles.[26] Let T(z) be **sum**(1, ..., z), the sum of the first z positive integers. Since the visualizer knows that the number of dots in each of the two triangular arrays is T(z), a disposition to infer in accord with the forms for {×} yields the belief that the number of dots in the two triangular arrays combined is $2 \times T(z)$. Since the two triangular arrays combined and the rectangular array are one and the same, and as the number of dots in the rectangular array is $z \times (z+1)$, we infer that $2 \times T(z) = z \times (z+1)$. Thence antecedent conceptual knowledge linking multiplication with division comes into play, namely, that for any positive integers h, j, and k,

if $h \times j = k$, then $j = k/h$.[27]

This yields the belief expressed by formula $T(z) = z \times (z+1)/2$, where it is understood that $T(z)$ dots can be arranged to form a triangular array, n dots in the nth column, etc. For reasons given above special provision must be made to verify the formula when z is 1. What has been described here is a route to the formula for all integers z greater than 1.

Conclusion

The creative heart of the discovery process lies in viewing a form in two ways at once. This was illustrated in Chapter 4 by the visual route to the theorem that the square whose vertices are midpoints of the sides of another square has half the area of this other square: a particular line segment was viewed as both a diagonal of one square and a side of another; a certain triangle was viewed both as half of one square and a quarter of another. It is illustrated again in this chapter. In the process of constructing in visual imagination the final array of dots, one accumulates information descriptive of the final image. "Viewing" the final array both as a rectangular array and as composed of two equal triangular arrays is a matter of selecting from the accumulated information two descriptions of the final form, a selection which may be prompted by the imagery experience and the problem to be solved. These descriptions are then used to make conceptually licensed inferences.

It is clear that this is a non-empirical mode of thinking in which the experience of visualizing plays an essential role, so again we have examples of synthetic *a priori* ways of reaching mathematical truths, but this time in the domain of number theory rather than geometry. In general terms the role of visualizing experience in these arithmetical cases can be described in the following way. The items of visualizing experience, the phenomenal images, are vehicles of information, and operations on those items are ways of accessing further information and combining it with already accessed information, so that conceptual resources can be applied to obtain something new. Provided one is vigilant for unintended exclusions, this kind of thinking can be reliable; it can be a way of discovering general theorems covering an infinity of cases. The extent of general arithmetical theorems discoverable this way is not clear. I would expect that all the theorems based on the ancient Greek "pebble" arguments fall into this class and indefinitely many like them. This does not take us very far into number theory, but it goes beyond common knowledge, and for that reason should not be regarded as insignificant.

Notes

1. $[(n/m) = \sqrt{2}] \Leftrightarrow [(n/m)^2 = 2] \Leftrightarrow [n^2/m^2 = 2] \Leftrightarrow [n^2 = 2m^2]$, where n, m are positive integers. The irrationality of the positive square root of 2 is said to have been discovered by a member of the Greek religious sect of the 5th century BC founded by Pythagoras. It was a dogma of the Pythagoreans that all phenomena were ordered according to integers or ratios of integers (according to Aristotle, *Metaphysics*, 985^{b}23-986^{a}26) and they knew that the positive square root of 2 is the length of the diagonal of a unit square, so this discovery would have been thought to contradict the dogma.
2. Kline (1972: 29-30).
3. I count as "visual experiences" both experiences of seeing and experiences of visualizing.
4. See Kosslyn (1983: chs. 3, 7); Kosslyn (1980: chs. 3, 5, 7). Kosslyn often uses scare quotes around "zoom in" and "scan".
5. Compare Tye (1993: 369). I take (1) from Tye. He gives a different second condition. Both second conditions assume that there is an analogue of spatial extension and location within the image field and that representing image parts have "locations" within that field. See Kosslyn (1983: 22–4) for an idea of the way an analogue might work.
6. Michael Tye brought this to my attention.
7. For further examples of picture indeterminacy see Block (1983: 651–61), Tye (1991: ch. 6), and Tye (1993: 356–71). Note that there are two kinds of indeterminacy: indeterminacy with respect to the presence or absence of a property (that would if present be visible to viewers) of a represented object; indeterminacy with respect to the particular property (properties) in a property spectrum that is represented

as instantiated. Examples of the first kind of indeterminacy need not be examples of the second kind.
8. This is a problem well known to beset reasoning with diagrams, referred to as *over-specificity* by Shimojima (2004b), in which Shimojima illustrates and analyses this and other use-relevant features of diagrams. Shimojima (2004a) is an abstract with bibliography for (2004b).
9. The limit may depend on the kind of array, and there may be individual variation.
10. Yet 1 is a triangular number. There is no contradiction here, as the definition of "triangular number" is arithmetical: the nth triangular number is the sum of integers from 1 to n.
11. There is a way of dealing with the source of these harmful exclusions that allows visualizing to deliver all of what was originally intended. In place of the visual properties of *square* of dots and *line* of dots one can use the following disjunctive properties:

*square of dots ≡ square of dots *or* single dot;
*line of dots ≡ line of dots *or* single dot.

The single dot is a limiting case of both a *square of dots and a *line of dots. As 1 dot can be arranged to form a *square of dots and a *line of dots, the visualized procedure can reveal that if a number of dots greater than 1 can be arranged to form a square, it can be arranged to form a square consisting of *either* 4 equal *squares of dots *or* 4 equal *squares of dots plus 4 equal *lines of dots plus a single dot. This holds for squares of 4 or 9 dots, as well as all larger ones. Clearly the even squares consist of 4 equal *squares. The following link between the visual and the arithmetical gets us to the originally intended theorem: For every positive integer k (including 1), k dots can be arranged to form a *square if and only if k is a square number.
12. I choose Kosslyn's theory because it is a pretty comprehensive theory that is well regarded among cognitive scientists. The theory of Kosslyn (1994) retains much of the theory of Kosslyn (1980). But there are some changes and some additions.
13. In Kosslyn (1994) activity in the visual buffer is identified with cell firings in the retinotopically organized visual areas of the occipital lobe. So an image in Kosslyn's later theory is a pattern of neural activity. An image in this theory cannot be identified with a phenomenal image, as

a pattern of neural activity is not subjectively discernible. As explained in Ch. 6 n. 34, this is not a concession to dualism about mental events.
14. A "category pattern" is Kosslyn's term for what I have been calling a category specification. Where I have needed to discuss category specifications at length I have replaced the word "pattern" to avoid the suggestion that a category specification consists in a visual template or visual prototype of the category, as opposed to a set of codes for visual features.
15. Kosslyn (1994) also gives an account of the cortical locality and arrangement of the processors in his theory of imagery. For a more recent view of the neural systems which realize these processors, see Thompson and Kosslyn (2000).
16. Kosslyn (1994: 292). See (1994: 291–5) for Kosslyn's theory of multi-part image generation.
17. Kosslyn (1994: 87–91; 291–2).
18. To visualize parts too small to be included in the original image such as the left thumbnail when no detail smaller than a thumb is represented, the attention window shifts and narrows down to surround a small part of the global image, perhaps the part representing the left hand. The spatiotopic mapping system now alters the image generating function (the mapping function in Kosslyn's terminology) from the pattern activation system to the visual buffer so that a pattern representing the left hand becomes the global image. A representation of the left thumbnail is now included in the image. As before, to increase resolution the attention window shifts and narrows down to the relevant part of the new global image.
19. Elements of a psychological image may stand in relations that are analogues of spatial relations. These analogues are order preserving in the sense that items $x_1 \ldots x_n$ are represented as standing in spatial relation R if and only if the representing elements $\mathbf{i}(x_1) \ldots \mathbf{i}(x_n)$ stand in the analogue of R. When I talk as though elements of an image have some spatial property or stand in some spatial relation, I intend the analogues. See Kosslyn (1983: 20–5).
20. Let me stress that this is just one possibility. A variant process of enlargement replaces the original configuration altogether but notes its place in the column-array. That location information would be encoded in the spatial relations system and would be available for getting the

attention window to shift and to narrow down to the relevant location, so that the original dots come back into view, just as starting with an image of a human body one gets an image of its left little-fingernail.

21. After writing this I came across Mateja Jamnik's outstanding work on schematic diagrammatic proofs, illustrated by analysis of a pebble proof of the proposition that for any positive integer n, $n^2 =$ the sum of the first n odd numbers, in Jamnik (2001). Her work is consonant with the approach I am proposing, though her ultimate concern is automated proof (as opposed to human discovery).

22. Explanatory notes about {**sum**}. (1) The union of disjoint sets si, $1 \leq i \leq z$, is simply the result of combining all the sets, so that x belongs to the union if and only if, for some i between 1 and z inclusive, x belongs to Si. (2) The letter "z" is a schematic variable for integers greater than 1. (3) For pairs of sets, {**sum**} coincides with a concept of binary cardinal addition {+} such that for all integers z greater than 1, **sum**$(k1, \ldots, kz) = (\ldots ((k1 + k2) + k3) \ldots + kz)$.

23. Generating a copy should be simpler than merely repeating the whole process by which the original was generated, but copying is not discussed in Kosslyn's expositions.

24. Motion-added transformations are to be contrasted with motion-encoded transformations. The latter occurs by activating a stored representation of a moving object, the motion having been encoded at the time of perception.

25. This mechanism for image transformation is new to the model of Kosslyn (1994). See Kosslyn (1994: 345–57), for differences between the new and old accounts.

26. For empirical data and discussion of this phenomenon see Finke, Pinker, and Farah (1989).

27. That this could be conceptual knowledge is argued in Giaquinto (1996).

9
Visual Thinking in Basic Analysis

Analysis is the general theory that underpins calculus. Calculus grew out of attempts to deal with quantitative physical problems which could not be solved by means of geometry and arithmetic alone. Many of these problems concern situations which are easy to visualize. In fact visual representations are so useful that most books on calculus are peppered with diagrams.[1] Yet visual thinking in analysis is often thought to be especially untrustworthy. There is no contradiction here. Visualizing may have various roles. For example, visual illustrations may facilitate comprehension of formulas or definitions; they can be reminders of counter-examples to plausible seeming claims; they can serve as stimuli, to spark an idea for a proof. These uses of visual thinking do not involve trusting it to deliver or preserve truth. So visualizing, or seeing a diagram in a particular way, may be useful when its role is merely to illustrate or to stimulate, but untrustworthy when used as a means of discovery.

This is a common view: visualizing in analysis, though heuristically useful, is not a means of discovery, let alone proof. But in other areas of mathematics, Euclidean geometry and arithmetic of the positive integers, the investigations of earlier chapters revealed that visualizing can be not merely a stimulus but also a means of discovery. Might some visualizing turn out to be a means of discovery in analysis too? Or is there something peculiar about analysis which bars visualizing except as provider of illustration and stimulus? Experts have sharply disagreed: Landau regarded diagrams in analysis as deceptive, while Littlewood thought that a diagram could provide proof of an analytic theorem. Given that they were talking about visualizable diagrams, they must have disagreed about the epistemic utility of visualizing as well as seeing diagrams.

What is the truth of the matter? This chapter approaches the question by focusing on some simple examples of visual routes to belief in theorems of basic real number analysis. First I present two examples and argue that they do not constitute paths of discovery. Then I consider some attempts at a general diagnosis of the obstacles to visual discovery in analysis. Finally, I investigate the possibility of overcoming the obstacles, again using a particular example.

Discovery of Rolle's Theorem by visualizing?

Co-ordinate geometry provides a useful way of representing functions on real numbers: a line, which may be curved, jagged, or interrupted, is drawn over calibrated horizontal and vertical axes to represent a function f, according to familiar conventions.[2] Using these conventions, a continuous function can be represented (if at all) only by an uninterrupted or gapless line, a line drawn without lifting the pencil tip off the paper, so to speak; and a differentiable function can be represented (if at all) only by an uninterrupted line which is also perceptually smooth, that is, unjagged, without sharp changes of direction.[3] Again using these conventions, we have (in some cases at least) a way of representing the value of the derivative of a function f at a point p, assuming that there is one. If the curve of f is perceptually smooth and with an upward but increasingly gentle slope in a neighbourhood of p, we can visualize a straight rigid rod rolling over the curve in that region, so that at each moment the rod touches exactly one point of the curve.[4] When it touches point p it represents the tangent of f at p, and the rod's inclination represents the derivative of f at p.

Do these possibilities of visual imagining provide a way of discovering Rolle's Theorem? Here is a statement of the theorem:

> If f is continuous on **[a, b]** and differentiable on **(a, b)** and $f(\mathbf{a}) = f(\mathbf{b})$, then for some x between **a** and **b**, $f'(x) = 0$.

Here it is understood that f is a function on the reals with derivative f' and that $\mathbf{a} < \mathbf{b}$.[5] If, for x between **a** and **b**, $f(x)$ never rises above or falls below $f(\mathbf{a})$, the curve of f over **[a, b]** is a horizontal line, as $f(\mathbf{a}) = f(\mathbf{b})$; and so for every point x between **a** and **b**, the tangent to f at x would have zero gradient, hence $f'(x) = 0$. So to establish the theorem it suffices to consider the cases in which $f(x)$ does rise above or fall below $f(\mathbf{a})$, for x somewhere between **a** and **b**.

Suppose then that $f(x)$ rises above $f(\mathbf{a})$, for x somewhere between **a** and **b**. Since the curve of f over [**a**, **b**] must be smooth and uninterrupted, the curve must rise smoothly somewhere over [**a**, **b**]. But then it must fall smoothly again in order to reach its endpoint $\langle \mathbf{b}, f(\mathbf{b})\rangle$, as $f(\mathbf{a}) = f(\mathbf{b})$. So the gradient of the tangent to the curve of f must change gradually from positive to negative at least once, as would happen if a segment of the curve had the shape of a path over a smooth hilltop. Now visualize a segment of the curve as it undergoes any one of these changes. We can visualize a straight line segment rolling over this part of the curve like a straight and rigid rod, so that this straight line segment—call it "the rod"—represents an interval of the tangent containing the point of contact with the curve. If we visualize this, the gradient of the rod will appear to change, falling (gradually, because the curve is smooth) from positive to negative; that is, it will appear initially inclined upward (from left to right), and then it will appear to become gradually less and less upwards inclined, eventually moving through a horizontal position to become downwards inclined. See Figure 9.1.

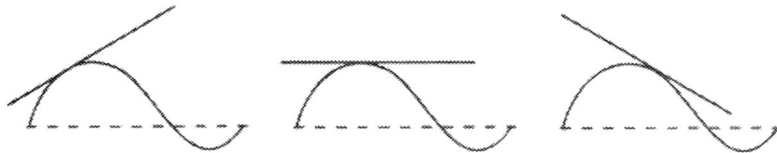

Figure 9.1

Visualizing in this way it seems that the tangent must at some point be horizontal, otherwise there would have been a sharp change in the direction of the curve. At this point then the gradient of the tangent must be zero and so $f'(x) = 0$, for some x between **a** and **b**.

There remains the case in which $f(x)$ falls below but never rises above $f(\mathbf{a})$. Here we visualize as described above, supposing the curve to be upside down, with the endpoint $\langle \mathbf{b}, f(\mathbf{b})\rangle$ to the left of the initial point $\langle \mathbf{a}, f(\mathbf{a})\rangle$. Making the obvious changes in the accompanying argument (such as interchanging "**a**" with "**b**", "rising" with "falling"), we reach the same conclusion.

Here is a way in which a person could come to believe Rolle's Theorem. Granted, it does not constitute a way of proving it. But there may be discovery without proof. So the question remains: if one came

166 VISUAL THINKING IN BASIC ANALYSIS

to believe Rolle's Theorem in the way described above, where visual imagination plays a crucial role, would one have genuinely discovered it? In the following section I argue not.

Objections

(a) The following preliminary objection will occur to anyone who is familiar with the standard proof of Rolle's Theorem.[6] From the supposition that the curve rises smoothly and falls again over [**a**, **b**] it was inferred that the gradient of the tangent to the curve must change gradually from positive to negative at least once, as would happen if a segment of the curve had the shape of a path over a smooth hilltop. The hilltop image is then relied on in the ensuing instructions to visualize. But this overlooks a serious question. How do we know that the hilltop image is not misleading? The hilltop curve represents a function whose values have an upper bound. How do we know that the values of any function satisfying the conditions of the theorem have an upper bound? One of those conditions is continuity, and it may be that every function unbounded on a closed interval that comes to mind is discontinuous. But what about functions unbounded on a closed interval that do not come to mind? It is not a trivial task to establish that every function continuous on a closed interval is bounded on that interval. So there is a gap in the line of thought which prevents it from being a route of genuine discovery.

(b) Set aside the last objection. That is, assume one knows that a function continuous on a closed interval [**a**, **b**] is bounded on [**a**, **b**]. Assume in addition that one knows that the function reaches its least upper bound and greatest lower bound on [**a**, **b**]. Is the smooth hilltop image still misleading? Yes. To see why, consider the following. A function has a curve that is (perceptually) smooth and at first appears to have a single peak; closer examination of the peak reveals a small shallow dip, so that now this top segment of the curve appears to have two mounds separated by the little dip; yet closer examination of the tops of the two mounds and the bottom of the dip reveals further fluctuation, each mound again having a dip but shallower than before, and the first dip having a gentle mound separating two very shallow dips. Now let this be replicated unendingly, so that increasing the scale uniformly reveals ever smaller dips in the new mounds and ever smaller mounds in the new dips, as illustrated in Figure 9.2.

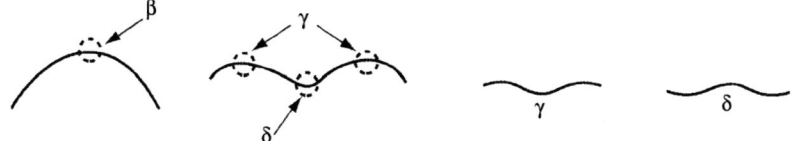

Figure 9.2

How in this situation can we visualize the movement of the tangent to the curve as it changes gradient from negative to positive? We cannot do so: over any interval in which the gradient changes from positive to negative it must make this change infinitely often, something which cannot be visualized. Nor could we rule out this kind of curve as impossible or irrelevant without further argument. A curve meeting the description of the previous paragraph would be uninterrupted and unjagged, and so the function it represents does not obviously violate the conditions of Rolle's Theorem. We would need to argue analytically that a curve of this description cannot represent a function that is differentiable everywhere in the interior of the interval that constitutes its domain.

(c) Let us restrict our attention to functions for which the hilltop image is appropriate, such as certain quadratics. Would the visual part of the thinking described in the previous section then be an acceptable way of stepping to the conclusion that the derivative of the function is zero at some point over [**a**, **b**]? Again, no. As before, let us call the line segment representing the tangent "the rod". If we reach Rolle's Theorem in the manner described above, we assume the following: if, in visualizing the rod roll over the curve, it appears to assume a horizontal position at some point, the function has a zero derivative at that point. But is this right? Can we not visualize the rod assuming a horizontal position as it rolled over a curve where there is *no* zero derivative? We can. Consider a function on the interval $[0, \pi/2]$ that rises exactly as the sine wave rises over $[0, \pi/4]$, but then falls exactly as the sine wave falls over $(3\pi/4, \pi]$. Though the curve of this function changes direction sharply, if we visualize a straight rigid rod rolling over the curve we can visualize it assuming a whole range of gradients, including the horizontal, when its point of contact is the sharp peak of the curve, $\langle \pi/4, \sin(\pi/4) \rangle$. Yet there is no point at which this function has a zero derivative. Thus visualizing a rod move through the horizontal orientation as it rolls over the curve of a function would not be a reliable indication that the function has a zero derivative.

There is a response to this objection. Rolle's Theorem concerns only those functions which have a derivative at every interior point of the interval on which it is defined,[7] so the curve of the function must be smooth, unlike the function just described. For the purpose at hand visualizing the moving tangent as a rigid rod rolling over the curve need only be a reliable indicator of the derivative's zeros for functions whose curve is smooth.

That response works only if the apparent gradient of the rod really is a reliable indicator of the value of the derivative. But for this, the apparent gradient of the visualized rod must be constrained by the derivative in the following sense: if the derivative were not zero the rod would not appear horizontal; if it were zero the rod would not appear inclined. It turns out that our visualizing propensity is not constrained in this way. In one study, students in a service level calculus course were presented with the graph of x^3 around the origin somewhat as in Figure 9.3. Most of the students stated correctly that there is just one tangent to the curve at the origin, but not even a fifth of them drew the tangent correctly.[8]

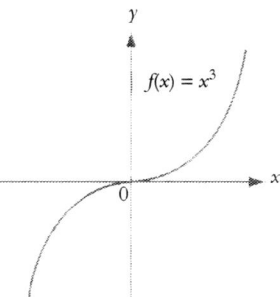

Figure 9.3

In this case the tangent to the curve at the origin does have a zero gradient, i.e. the derivative of x^3 is zero when x is zero, but people show a tendency to imagine a tangent with a small positive gradient. Let $\alpha°$ be the angle between the horizontal and the imagined tangent. How would the same students have viewed a curve obtained by clockwise rotation through $\alpha°$ of the graph of x^3 about the origin? This was not tested, as far as I know. But it is quite possible that they would have taken the tangent of *that* curve at the origin to be horizontal. That would be a case of visualizing a curve as having a horizontal tangent where it is in fact inclined, that is, where the curve represents a point at which the function has a negative derivative, not a zero derivative. Further evidence of the unreliability of our visual capacity to detect zero derivatives by this method is that we can visualize, even see, more than one rod—rods of different gradients—as tangent to a curve at a given point. This may be due to the impossibility of visualizing or seeing a rod as touching a curve at just a single point as distinct from a very small segment of the curve around the point.

For these reasons the way of reaching Rolle's Theorem described in the previous section should not be regarded as an epistemically acceptable way of achieving belief in the theorem. The visual means are unreliable; so the visual thinking in this case is not a way of making a genuine discovery.

Discovery of Bolzano's Theorem by visualizing?

Perhaps visualizing could turn out to be trustworthy as a means of discovering analytic theorems when visualizing is not used to determine a derivative. After all, the notion of the tangent to a curve does not arise naturally from experience except in the case of an arc of a circle; even some of the familiar applications of the notion of derivative, such as velocity of a moving body *at a point of time* are intuitively puzzling. So let us consider a visual way of coming to believe a theorem which does not involve derivatives, Bolzano's Theorem:[9]

> If f is continuous on [**a**, **b**] and $f(\mathbf{a})$ and $f(\mathbf{b})$ have opposite signs, $f(x) = 0$ for some x between **a** and **b**.

Bolzano's Theorem is much simpler than Rolle's Theorem. It does not involve the concept of the derivative of a function, and there is a fairly straightforward way of visualizing instances of the situation described in the antecedent condition. Imagine a calibrated horizontal line, the x-axis. We take it that the curve of the function meets the x-axis at z if and only if $f(z) = 0$. Now let **a** and **b** be distinct, let $\langle \mathbf{a}, f(\mathbf{a}) \rangle$ lie on one side of the x-axis and $\langle \mathbf{b}, f(\mathbf{b}) \rangle$ on the other. For definiteness let $f(\mathbf{a})$ be negative, so that the left endpoint lies below the x-axis and the right endpoint above. (One can repeat the process to be described imagining the left endpoint above and the right endpoint below.) As f is continuous, the curve of f

must be an uninterrupted line. The *x*-axis is infinite in both directions; there is no way of going round it in order to reach a point above from a point below. Now visualize tracing out an uninterrupted path, as circuitous as that of Moses in the desert if you like, from $\langle \mathbf{a}, f(\mathbf{a}) \rangle$ below the *x*-axis to $\langle \mathbf{b}, f(\mathbf{b}) \rangle$ above. The *x*-axis will not split open like the Red Sea, and there is no jumping over or tunnelling under: the curve stays in the plane. Visualizing the matter thus, it seems clear that the curve must meet the *x*-axis.

This is no proof of the theorem.[10] But is it not a legitimate way of becoming convinced? We take every continuous function to have an uninterrupted curve, i.e. a curve without gaps, and we take a function whose curve meets the *x*-axis to have a zero value. These assumptions link the analytic concepts with the visual. What visualization contributes is the conviction that every curve without gaps from a point below the *x*-axis to a point above actually meets the axis. Bolzano's Theorem is inferred from this and the assumptions linking analytic with visual concepts. Spelled out the inference runs thus:

1. Any continuous function f on $[\mathbf{a}, \mathbf{b}]$ such that $f(\mathbf{a}) < 0 < f(\mathbf{b})$ has an uninterrupted curve from a point below the *x*-axis to a point above. (LINK 1)
2. Any uninterrupted curve from a point below the *x*-axis to a point above meets the axis. (FROM THE DIAGRAM)
3. Any function whose curve meets the *x*-axis has a zero value. (LINK 2)
4. Therefore any continuous function f on $[\mathbf{a}, \mathbf{b}]$ such that $f(\mathbf{a}) < 0 < f(\mathbf{b})$ has a zero value.

What is there to object to in this?

Objections

Despite their apparent obviousness, I think that both link assumptions, LINK 1 and LINK 2, are false.

Consider first LINK 2: any function whose curve meets the *x*-axis has a zero value. Here the idea of a *meeting* is visual, not analytic: a curve meets the *x*-axis when it visually appears to meet it. The assumption links this visual idea with the analytic idea of a function's having a zero value. How can the assumption be false? Consider the following functions defined on $[-1, 1]$ or a subset of it for comparison.

$g(x) = x$.
$h(x) = x$, for $x \neq 0$; $h(0)$ undefined.
$j(x) = x$, for x in $[-1, 0)$; $j(x) = x + \varepsilon$, for x in $[0, 1]$, where for any given unit distance, ε is so chosen as to be less than a Planck length.

The curves of these functions are visually indistinguishable. The function g both meets the x-axis and has a zero value at $x = 0$. The curve of the function h is indistinguishable from the curve of g, even though h has no zero value. One might think that the curve of h has a visible one-point gap at $x = 0$. But this is a mistake. If it did have a visible gap, the two sections of the curve either side of the gap would be visibly separated by some space, and so there would appear to be a positive interval between the two sections; they could not appear to be separated by just one point, as a point has zero breadth. And there may be an interval that is simply too small to be a visual gap, as exemplified by j. So all of these functions meet the x-axis but only one of them has a zero value.

Another way to see the point is to consider the following two functions:

$k(x) = x^2 - 2$, for x in $[1, 2]$.
$m(x) = x^2 - 2$, for *rational* x in $[1, 2]$, otherwise undefined.

The curves of these functions are visually indistinguishable. For any rational there are rationals arbitrarily close to it on both sides; so between any two rationals there are infinitely many others. This means that the curve of m will have no visible gaps; its curve is visually indistinguishable from the curve of k. But m has no zero value as there is no rational x for which $x^2 - 2 = 0$, even though it is clear that $m(1)$ is negative and $m(2)$ is positive.[11]

These considerations refute LINK 2, the assumption linking the visual idea of a curve's meeting the x-axis with the analytic idea of a function's having a zero value. This undermines the possibility of genuinely discovering Bolzano's Theorem in the visual manner described above. But there is a reply: the counter-examples to LINK 2 that have been given are all discontinuous functions, whereas Bolzano's Theorem is about continuous functions; so instead of LINK 2 we can use this modification of it:

Any *continuous* function whose curve meets the x-axis has a zero value.

Wrong again. Let g be a continuous function defined only on the open interval (**a**, **b**) and let it be an increasing function with $g($**a**$)$ negative such that the limit of $g(x)$ as x approaches **b** is 0. Then g will be a continuous function whose curve meets the x-axis with no zero value. Its curve meets the x-axis because there is no gap, visible or otherwise, between the tail of the curve and the x-axis.

To block this we could make a further modification: any function continuous *on a closed interval* whose curve meets the x-axis in that interval has a zero value. This may be true. But is this a claim that we can reliably assume without further argument? Clearly not. We arrived at it by successive modifications designed to avoid earlier falsehoods. We have no positive basis for believing it, beyond the original fallacious inclination to infer the analytic condition (having a zero value) from the visual condition (meeting the horizontal axis). Reaching it the way we just did places it beyond the class of beliefs on which one could base a genuine discovery.

Let us turn to the first link assumption, LINK 1:

> Any continuous function f on [**a**, **b**] such that $f(\mathbf{a}) < 0 < f(\mathbf{b})$ has an uninterrupted curve from a point below the x-axis to a point above.

Without doing any damage to the argument in which this assumption is used we can factor it into a conjunction as follows.

> Any function f continuous on [**a**, **b**] has an uninterrupted curve and if $f(\mathbf{a}) < 0 < f(\mathbf{b})$ that curve runs from a point below the x-axis to a point above.

Let us restrict attention to functions defined on a closed interval. The problem with this is not that some continuous functions have interrupted curves, but that some of them have no curve at all. Here the term "curve" is being used for graphs of functions visualizable as a line (not necessarily straight), and the terms "interrupted" and "uninterrupted" signify visible and visualizable properties of them. Any function defined on the closed unit interval that is continuous but nowhere differentiable is an example of a continuous function without a curve. We do have a visual way of reducing the puzzling character of such a function: we visualize first a curve with sharp peaks and sharp valleys; then we imagine zooming-in on an apparently smooth part between a peak and a valley, only to find that this part itself contains sharp peaks and valleys; then we imagine zooming-in on an apparently smooth part of this, again finding that it is not smooth but contains sharp peaks and valleys; and we suppose that this is repeatable without end. However, this is not a way of visualizing the curve of a function. At every stage the curve visualized has a smooth part and so cannot be the curve of a nowhere differentiable function. The curve of a continuous function which is non-differentiable at a point makes a sharp turn at that point, and a curve consisting of sharp turns at every point, without any smooth segments between sharp points, is unvisualizable.[12]

So as in the visual route to belief in Rolle's Theorem, there is an illegitimate generalization. A claim about all functions continuous on a closed interval is inferred from a belief about all such functions that have a visualizable curve. But in this case the fault lies not in using a particular type of uninterrupted curve (the smooth hilltop) as a representative image, but in using the image of an uninterrupted curve to represent continuity: some discontinuous functions have uninterrupted curves and some continuous functions do not have curves at all. In sum, both LINK 1 and LINK 2, the assumptions linking the visual with the analytic, are false. For this reason, the visual route to Bolzano's Theorem described above is not a path of genuine discovery.[13]

Visualizing and limits

Can we give a uniform diagnosis of epistemic failures of visualizing in analysis? The considerations of previous sections have shown that neither continuity nor having a zero derivative have completely adequate visual counterparts. Perhaps this is because these properties, like all or most fundamental analytic properties, are about what happens at the limit of an infinite process. Should we say, then, that visualizing becomes unreliable whenever it is used to discover the existence or nature of the limit of some infinite process?

There is no doubt that our visual intuitions about what happens at the limit of an infinite process sometimes lead us astray. My favourite example is this.[14] Consider the following sequence: first, a semicircle on a diameter; the second curve is formed from the semicircle over the left half and the semicircle under the right half; if a curve consists of 2^n semicircles alternating above and below the diameter, the next curve results from dividing the original segment into 2^{n+1} equal parts and forming the semicircles on each of these parts, alternately over and under. The first three curves are illustrated in Figure 9.4. Obviously the curves get closer and closer to the diameter. For large n, the nth curve will

174 VISUAL THINKING IN BASIC ANALYSIS

be a wiggly line hugging the diameter, becoming barely distinguishable from the diameter. The curve lengths too seem to approach that of the diameter, so that the limit of the curve lengths will be the length of the diameter.

But this is wrong. The length of all curves in the sequence is the same. If the diameter of the first semicircle has length d, that semicircle, which is the first curve, has length π.d/2. To get the next curve after any given curve, we double the number of semicircles and halve their diameters. So the lengths of the curves are as follows.

Curve 2: $2(\pi.d/4) = \pi.d/2$.
Curve 3: $4(\pi.d/8) = \pi.d/2$.

And in general

Curve n : $2^{n-1}(\pi.d/2^n) = \pi.d/2$.

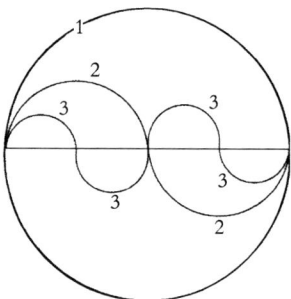

Figure 9.4

This lends credence to the idea that visualizing is never reliable when used to discover the nature of the limit of an infinite process. This may be the correct explanation of the famous deceptiveness of visualizing in analysis, since its basic concepts all involve limits of infinite processes. Even so, it remains possible that visualizing can play a non-redundant role in discovery in analysis. I now want to explore this possibility by examining a particular example.

Consider the following visual way of coming to believe the theorem that the limit of the series $1/2 + 1/2^2 + 1/2^3 + \ldots + 1/2^n + \ldots = 1$. Imagine a square; now imagine dividing it in half by a vertical line; now imagine writing "1/2" in the left half and dividing the right half into squares by a horizontal line; now imagine writing "$1/2^2$" in the lower right quarter and halving the top right quarter by a vertical; now write "$1/2^3$" in the left half of this quarter and divide its right half by a horizontal; and so on.[15] Figure 9.5 illustrates the first few steps.

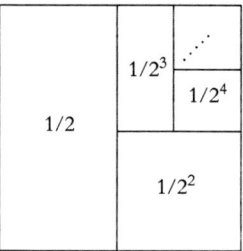

Figure 9.5

The following two propositions seem clear.

(α) At no stage of the process do the areas of the parts marked with a figure fill out the square; so taking the area of the square as unit, 1 is an upper bound of the series.

(β) No region of the top right-hand corner is so small that it will not eventually be eaten into by the marked parts; so 1 is the least upper bound of the sequence.

The theorem can be inferred from (α) and (β), given that a monotonic increasing sequence bounded above converges to its least upper bound. But are there epistemically acceptable visual routes to (α) and (β)?

In coming to believe that 1 is an upper bound of the series, the crucial thought was (α): at no stage of the process do the parts marked with a "$1/2^n$" fill out the whole of the square. How could visual thinking lead to this thought? One way is this. Visualizing the first few steps in the process leads us to grasp the common form of each step, part of which can be described thus: one divides the unmarked rectangular area into rectangular halves, marks one of these halves and leaves the other half unmarked (to be divided at the next step). From this the crucial thought is immediate: at each stage there remains an unmarked rectangular part of the square. So the total area of the marked parts at each stage is less than the area of the square. So each sum $1/2 + \ldots + 1/2^n$ is less than 1.

176 VISUAL THINKING IN BASIC ANALYSIS

This visual thinking appears to me to be a reliable way of reaching (α). Is there a reliable visual way of reaching (β), the claim that no region of the square remains at every stage disjoint from the marked parts? This is more difficult than for (α), but here is a possibility. First, the visualizing may lead us to notice that every second step leaves an unmarked square in the top right corner; and that for even n the unmarked square in the top right corner is of size $1/2^n$ just when the process has gone exactly n steps. Let us name and display this.

(γ) For even n the unmarked square in the top right corner is of size $1/2^n$ if and only if the process has gone exactly n steps.

We may come to realize this by grasping the form of the process. Now suppose that there remains a region R, perhaps very small, in the top right-hand corner not overlapped by marked parts, a *free* region. Without loss of generality we can assume that R is convex.[16] Region R will contain the top right vertex and, within a sufficiently small radius from the top right vertex, a quarter disc. Then there is a free *square* region within that quarter disc in the top right corner, a square region whose diagonal is a radius of the quarter disc. Let the area of this free square region be **r**. As no real area is infinitesimal we can assume that for some even positive integer n the whole unit square can be covered by n non-overlapping regions of size **r**. Let **n** be such an n. Then $\mathbf{nr} \geq 1$; so $\mathbf{r} \geq 1/\mathbf{n}$. In other words the free square of size **r** in the top right corner is at least as large as a square of size $1/\mathbf{n}$; so it will entirely contain the square area in the top right corner of size $1/2^n$. Bearing in mind proposition (γ) and the fact that **n** is even, this means that after $\mathbf{n}+2$ steps region R in the top right corner will be overlapped by marked parts; hence (β).

This thinking is far from entirely visual; but some of the non-visual parts refer to the visual and so cannot stand alone. Here then is a route to a belief about the limit of an infinite process, in which visual thinking plays a non-superfluous part. It seems reliable, though I do not claim to have established that it is. Until we find some sign of unreliability here, it remains open that visualizing can, in some restricted cases, play a non-superfluous role in discovering the existence and value of the limit of an infinite process.

Another example raises the possibility that genuine discoveries can be made by means of thinking that involves visualizing uninterrupted curves in reaching a conclusion about a restricted class of continuous functions. Littlewood noted in his brief "Postscript on Pictures" that a heavy warning used to be given against the use of pictures as lacking necessary rigour. Not so, he claimed, in contrast to Landau; and he went on to illustrate.[17] In his own words:

> One of the best pictorial arguments is a proof of the "fixed point theorem" in one dimension: *Let $f(x)$ be continuous and increasing in $0 \leq x \leq 1$, with values satisfying $0 \leq f(x) \leq 1$, and let $f_2(x) = f\{f(x)\}, f_n(x) = f\{f_{n-1}(x)\}$. Then under iteration of f every point is either a fixed point, or else converges to a fixed point.* For the professional the only proof needed is Fig. [9.6]

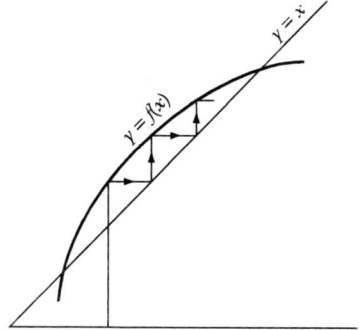

Figure 9.6

I would not call this a proof, even for the professional; a proof is a type of thinking which can be set out and publicly checked for validity (length and complexity permitting). But it may be possible to substantiate a weaker claim than Littlewood's: seeing or visualizing the diagram can initiate a partly visual path to discovery of the fixed point theorem. In Appendix 9.1 I set out a partly visual path to the theorem. Whether that path is a path of discovery depends on whether certain restricted propositions linking visual conditions with analytic conditions are true and known to be true by the thinker. Given the results of the earlier examination of a visual route to belief in Bolzano's Theorem, we have strong reason to be cautious about visual-analytic links. Only with considerable experience, perhaps only with the experience of "the professional" as Littlewood says, can one reliably tell true links from false.

A diagnosis

Why does visualizing so often fail as a means of discovering (even simple) general theorems in analysis, when in Euclidean geometry the reverse is true? There is a difference between basic geometric concepts and basic analytic concepts, a difference that must figure in the explanation of the relative impotence of visual thinking in analysis. As shown in Chapter 2, a basic concept for Euclidean geometric squares can be obtained by a modification of a perceptual concept for squares. Roughly speaking, to be an instance of that geometric concept for squares[18] is to be a *perfect* instance of the perceptual concept. Thus there is a tight intrinsic link between the perceptual concept and the geometrical concept. Moreover, applying the geometric concept involves thinking of its instances in terms of the appearance of a perceptually perfect square. This link with the perceptual explains why having the geometric concept entails having belief-forming dispositions that can be triggered by perceptual experience. We can exploit this link, as was illustrated in Chapter 3, to make geometrical discoveries by visual means.

The same applies to other basic concepts of Euclidean geometry, such as those for line segments and for circles. An edge of a coin can be an instance of a perceptual concept of circle, and the edge of a CD can be a still better instance of that concept, where the comparative is to be understood perceptually; and a perfect instance of that concept would be an instance of a geometric concept of circle. As the relevant geometric concepts are thus intrinsically linked to the perceptual, having them entails having belief-forming dispositions that can be triggered by perceptual experience. We do not need to employ assumptions linking the perceptual with the geometric.

But analytic concepts are not intrinsically linked to perceptual concepts. Consider the analytic concept of continuous function. We need only recall the definition in order to see how far removed it is from any perceptual concept of continuity.

> A function f is *continuous* if and only if it is continuous at every member p of its domain.
>
> For any member p of the domain of f, f is *continuous at p* if and only if for all $\varepsilon > 0$ there is a $\delta > 0$ such that if $|x - p| < \delta$, then $|f(x) - f(p)| < \varepsilon$.[19]

Possession of this concept of continuity does not bring with it belief-forming dispositions that can be triggered by perceptual or, more specifically, visual experience. In order then to employ visual means to arrive at analytic beliefs we have to import assumptions that link the perceptual with the analytic.

The assumption we tend to use in this case links the idea of a continuous function with the idea of an uninterrupted curve of a certain sort.[20] But the discussion earlier in this chapter shows that the set of functions representable by curves of that kind includes some discontinuous functions and excludes some continuous ones; so the visual and the analytic concepts mismatch in a way which prevents reliable inference to general conclusions about analytically continuous functions from premises about functions with visually continuous curves. This is not surprising given the history of the subject: the analytic concept of continuity would surely not have been developed had there been a visual concept that could be refined so as to make it adequate for mathematical purposes.

Similarly, the set of differentiable functions does not coincide with the set of functions representable by a perceptually smooth curve, as some differentiable functions are not visualizable as curves. This is not just a peculiarity of the particular properties of functions we have mentioned. The analytic concept of *function* itself is not co-extensive with any concept of a function as, or as representable by, a certain sort of curve. As mentioned earlier, there are analytic functions with no representing curve. What about the analytic concept of *integral*? Is it not co-extensive with a concept of the area under a curve? Not if some functions with an integral lack a visualizable curve. Setting aside this problem, we have to be careful about the visual concept of *area*. For arbitrary plane closed figures we have a visual concept of the *enclosed region*, and we have a visual concept of *regional size* which is vague and may even allow the possibility that there are two plane regions incomparable with respect to regional size. I am aware of no visual way of qualifying the vague concept to obtain a precise concept, one which, for a given unit square, assigns to each region a size which stands in some precise ratio to the size of the unit square. Nothing less precise will do for analysis. So the concept of integral cannot be characterized by means of the visual idea of the area under the curve.

To summarize: for no given basic analytic property of functions is there a visual property of curves such that the class of functions with the analytic property coincides with the class of functions whose graph has the visual property. This may be the main reason why, though visual routes to conclusions about all instances of basic geometrical concepts are often reliable, visual arguments for conclusions about all continuous/differentiable/integrable functions are very rarely so. This is not at all to suggest that when introducing the analytic concepts of *function, continuity, differentiability*, etc. definitions should not be accompanied by visual illustrations. Since the basic analytic concepts were developed to overcome shortcomings of perceptual concepts applied to physical situations, often visualizable situations about motion, it is not surprising that diagrams representing aspects of those situations can be used to illustrate applications. But the analytic concepts not only provided us with the means for clear and reliable thinking about those visualizable situations that were the original focus of concern; they also extended the domain of situations that can be treated uniformly way beyond what is visualizable.

The utility of visualizing in analysis

I have argued that visual arguments for analytic conclusions are rarely reliable, hence rarely are they paths of discovery. At the same time, visualization can have an important role in analytic discovery, by providing the idea for a proof. This is a way in which visualizing could be important to a working analyst such as Littlewood without having the epistemic role that he assigned to it in his example: we distinguish between a proof and an idea for a proof. Here is a simple illustration, one that might occur to a beginner. Consider the following route to the Intermediate Value Theorem from Bolzano's Theorem (which is sometimes known as the Intermediate Zero Theorem). The Intermediate Value Theorem is this:

> Let g be a continuous function over [**a**, **b**] such that $g(\mathbf{a}) \neq g(\mathbf{b})$. Then g takes every value between $g(\mathbf{a})$ and $g(\mathbf{b})$.

First imagine a curve running without interruption over a closed interval [**a**, **b**] with its endpoints at different heights, $g(\mathbf{a})$ and $g(\mathbf{b})$. Consider any height k (positive or negative) between the heights of the endpoints. Imagine the whole curve moved $-k$ units vertically. Say for definiteness that k is positive and that $g(\mathbf{a}) < k < g(\mathbf{b})$. Then we visualize the curve move downwards by k units; the result is a curve which appears to have its left endpoint below the x-axis and its right endpoint above. See Figure 9.7. The x-axis of course is the line "$y = 0$".

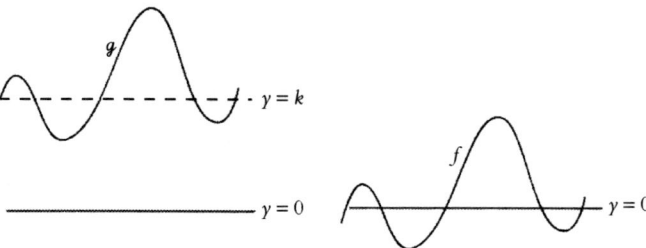

Figure 9.7

Assuming that mere translation preserves continuity, the newly positioned curve is the curve of a function which satisfies the conditions of Bolzano's Theorem, and so its curve meets the x-axis at some point **c** between **a** and **b**. If we imagine the whole curve moving back up again by k units, it is clear that the height of the restored curve at **c** is k, i.e. $g(\mathbf{c}) = k$. Since this procedure does not depend on which k between $g(\mathbf{a})$ and $g(\mathbf{b})$ is chosen, we can conclude that g takes every positive value between $g(\mathbf{a})$ and $g(\mathbf{b})$.

This is not an epistemically acceptable way of reaching the Intermediate Value Theorem: not every continuous function has a visualizable curve, so the visualizing does not present a form of thinking which in principle applies to all continuous functions. However, it obviously does provide the idea for an analytic proof. Imagining the curve of g move downward by k units suggests defining a function

$$f(x) = g(x) - k.$$

As $g(\mathbf{a}) < k < g(\mathbf{b})$ by hypothesis, $f(\mathbf{a}) < 0 < f(\mathbf{b})$; and f is continuous (because the difference of two continuous functions is continuous). So we can apply Bolzano's Theorem to f: $0 = f(\mathbf{c})$ for some **c** between **a** and **b**. Then imagining the curve move back up by k units suggests adding k to get:

$$k = f(\mathbf{c}) + k.$$

Then from the definition of f,

$$f(\mathbf{c}) + k = g(\mathbf{c}).$$

Hence for some **c** between **a** and **b**, $k = g(\mathbf{c})$, as required.

Notice that the analytic proof here is not just a translation of the visual thinking, because there is an illegitimate generalization involved when we move from a thought about all continuous functions with an uninterrupted curve to all continuous functions whatsoever. But the visual thinking clearly provides the idea for the proof.

Visual thinking is also immensely useful in other ways when learning calculus and analysis. This is shown by frequent use of diagrams in teaching and texts introducing calculus. I doubt that there is a soul who could have made much headway in calculus without any use of diagrams. In complex analysis too visual thinking can be extremely helpful, increasing both understanding and aesthetic pleasure.[21]

Here are some of the many ways in which diagrams and visual imagery can help students.

(1) Visual illustrations of instances of an analytic concept can strengthen one's grasp of the concept. Visual representations of simple instances, prototypes, seem to serve as organizers of thought, perhaps as iconic labels for collections of information involving the concept (e.g. definitions, frequently encountered instances) and perhaps as analogues for unusual instances. Errors may occur when the extension of the concept is unconsciously assumed to be determined by visual similarity to the prototype(s) rather than by the definition, but this is avoidable. The same goes for geometry.[22]

(2) When a function $f(x)$ has a visualizable curve, visualizing the curve can help us grasp and think about symbolically presented operations on the function[23] e.g.

$$f(x) + k, \quad f(x+k), \quad |f(x)|, \quad f(|x|), \quad f^{-1}(x),$$

and operations which cannot be grasped pointwise, such as:

$$I(x) = \int_0^x g(t)dt.$$

(3) It can help us to locate roughly values for given inputs, to grasp the type of a discontinuity, to find roots, and so on.

(4) It can help us avoid certain kinds of mistake. Researchers in mathematics education found that when students are given the task

Find the derivative of the function
$$f(x) = sin(x) \quad \text{if } x \neq 0$$
$$= 1 \quad \text{if } x = 0,$$

a common response is

$$f'(x) = \cos(x) \text{ if } x \neq 0, \text{ and } f'(x) = 0 \text{ if } x = 0.^{24}$$

The mistake arises from treating the two parts of the definition of f as though they defined separate functions; visualizing the curve of the function around the origin may reduce the likelihood of making this error.

(5) It can help reduce mystery, especially with the aid of computer imagery. A good example is given by Deborah Hallett, co-director of a calculus reform project. She writes as follows about the introduction of the exponential function $\mathbf{e}^x$:

> The traditional introduction of $\mathbf{e}^x$ as the inverse of [the log function] $\ln(x)$, which is itself defined as an integral, is usually lost on students who don't like logs and are not yet comfortable with integrals. As a result, the fact we'd most like them to know—that the derivative of $\mathbf{e}^x$ is itself—gets completely buried by the feeling that exponentials are impossibly mysterious. Just why anyone should call a logarithm "natural" is beyond them. However, with [computer graphing] technology it is easy to show students graphically just why $\mathbf{e}$ is so important. If the graphs of a^x and $d(a^x)/dx$ are displayed together, students can experiment, and with a little prodding, they "discover" $\mathbf{e}$ as the value of a making the graphs coincide.[25]

The utility of visualizing in these roles and maybe others has in the past been undervalued through a misdirected zeal for rigour in analysis, though this is rare today. There is no doubt that we can be easily fooled in analysis by our visual intuitions. The mathematicians who at the end of the nineteenth century transformed analysis into a numerical symbolic system gave us a safeguard against visual deception. Nonetheless the experience of researchers, teachers, and students of analysis attests to the usefulness of visual thinking in the subject, when combined with the proper "epsilon-delta" methods of analytic proof.

Conclusion

On the basis of the foregoing it is reasonable to expect that only in a very restricted range of cases can visual thinking be a mode of discovery in analysis. Visual thinking is generally unreliable as a means of reaching conclusions in analysis, largely because analytic concepts, so far from being faithful counterparts of visual concepts, apply to situations many of which are unvisualizable or situations which are visualizable but confound expectations born of visuo-spatial experience. The weakness of visual thinking as an epistemic tool in analysis is totally consistent with its great heuristic value. Many visual arguments that seem to reveal the truth of some proposition of analysis, that we might even be tempted to regard as a proof, in fact only provide the idea for a proof. But in this way such arguments can be very useful. With more experience a path from a visually provided idea to an analytic proof becomes easier to find, making visual arguments especially valuable to experts. This helps explain the positive attitude to visual thinking in analysis expressed by Littlewood and other working mathematicians. And there are many ways in which visual thinking can increase a student's understanding and technical expertise, none of which presuppose that visual thinking is a mode of discovery or proof. All this underlines one of the themes of this book: visual thinking has many distinct roles in mathematics.

Notes

1. Moshé Machover brought to my attention a notable exception: Landau (1934). It has no diagram, and no geometrical application.
2. The ordered pair $\langle x, f(x) \rangle$ is represented by the meeting point of the vertical line through x on the horizontal axis and the horizontal line through $f(x)$ on the vertical axis. If $x > 0$, the point lies to the right of the horizontal axis; if $x < 0$, the point lies to the left. If $f(x) > 0$, the point lies above the horizontal; if $f(x) > 0$, the point lies below. The unit of vertical distance is that between the 0-mark and the 1-mark on the vertical axis; similarly for horizontal in place of vertical. In many cases the set of points $\langle x, f(x) \rangle$ (where x is in the domain of f) is representable by a visualizable curve.
3. "Line" here does not imply straightness. James Brown reminded me that "smooth" has a technical meaning in analysis: a function is said to be *smooth* when it belongs to C^∞, i.e. when it has derivatives of every order. I am not using "smooth" in the analytical sense here.
4. Here and henceforth I use the term "curve" for a perceptible line rather than in the analytic sense of a continuous mapping of a closed interval into $\mathbb{R}^n$.

5. [a, b] is the set of x such that $\mathbf{a} \leq x \leq \mathbf{b}$; (a, b) is the set of x such that $\mathbf{a} < x < \mathbf{b}$.
6. Sueli Costa brought this to my attention.
7. If the function is defined on [a, b] the interior points are just the members of (a, b).
8. Vinner (1982).
9. Bolzano actually proved a more general theorem: If f and g are continuous on [a, b] and $f(a) < g(a)$ but $f(b) > g(b)$, for some x in (a, b) $f(x) = g(x)$. The special case where g is a constant function gives us what is known as the Intermediate Value Theorem. If the constant is 0 we get the theorem that I am calling Bolzano's Theorem, also known as the Intermediate Zero Theorem. See Bolzano (1817).
10. James Brown holds that the picture in this case does prove the theorem. See Brown (1997: 161–6) and Brown (1999: 25–30). The visual argument is not a proof of the theorem for the same reasons that it is not a means of discovering the theorem.
11. This example comes from David Tall's excellent article, Tall (1991).
12. Another example of a function that cannot be completely visualized is $x.\sin(1/x)$: its frequency increases without limit as x approaches 0. This and other examples are discussed in Hahn (1933).
13. Moshé Machover points out that if this were an acceptable way of arriving at Bolzano's Theorem, the corresponding visual way of arriving at Jordan's Curve Theorem would also be acceptable—which is not at all plausible.
14. I first learnt of this example from Sueli Costa.
15. David Gooding brought this example to my attention.
16. This is justified by the fact every Jordan region (with rectifiable curve) includes a convex set. See Theorem 10–42, Apostol (1957: 287–8).
17. Littlewood (1953: 54, 55). This was brought to my attention by Timothy Smiley.
18. There is more than one concept for Euclidean squares. Here I refer to the concept {**perfect square**} specified in Ch. 2. I expect that other concepts for Euclidean squares are related to this one by more than co-extensiveness.
19. We can take the domain and the range of the function to be included in the set of real numbers. For functions whose domain and range are

included in any metric spaces we can use the same definition (with indicators of the distance functions for those spaces).
20. These would have to be curves that do not double back on themselves, in order to capture the fact that a function cannot have more than one value for each member of its domain.
21. A tour de force in this regard is Tristan Needham's textbook *Visual Complex Analysis*, Needham (1997). I should also mention the classic of Hilbert and Cohn-Vossen (1952).
22. See Hershkowitz (1987).
23. See Eisenberg (1991).
24. Harel and Kaput (1991).
25. Hallett (1991).

Appendix 9.1 A Visual Path to a Fixed Point Theorem

Littlewood (1953) says the following.

> One of the best pictorial arguments is a proof of the "fixed point theorem" in one dimension:
>
> Let $f(x)$ be continuous and increasing in $0 \leq x \leq 1$, with values satisfying $0 \leq f(x) \leq 1$, and let $f_2(x) = f\{f(x)\}$, $f_n(x) = f\{f_{n-1}(x)\}$. Then under iteration of f every point is either a fixed point, or else converges to a fixed point.
>
> For the professional the only proof needed is Fig. [A9.1]

I reproduce Littlewood's figure here:

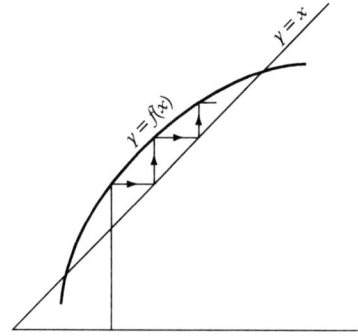

Figure A9.1

What follows is an attempt to map out a visual path to the fixed point theorem stated by Littlewood. To get going we need a restricted link assumption:

> If f is a function whose domain is a closed interval I and f is continuous and monotone increasing on I and whose range is included in I, f is representable by a curve.

Recall that z is a fixed point of f just when $f(z) = z$. Here is a second link assumption.

> If f is representable by a curve, z is a fixed point of f only if the curve of f meets the diagonal line "$y = x$" at $\langle z, z \rangle$.

The next claim is that if for some v in $[0, 1]$ $f(v) > v$, some z greater than v must be a fixed point of f, i.e. if the curve at v is above the diagonal, somewhere further to the right it must meet the diagonal. Otherwise f could not be defined for all non-negative x such that $x \leq 1$ without violating the condition that for all x less than or equal to 1, $f(x) \leq 1$. To see this, imagine the curve to the right of $\langle v, f(v) \rangle$ remaining above the diagonal: the curve at the point $\langle 1, f(1) \rangle$ would be above the point $\langle 1, 1 \rangle$, so $f(1)$ would be greater than 1, thus violating the condition. So if $f(v) > v$, the curve meets the diagonal somewhere over $(v, 1]$. Similarly, if $f(v) < v$, the curve meets the diagonal somewhere over $[0, v)$.

Littlewood's diagram illustrates what happens when the curve is above the diagonal. The small horizontal arrow from a point $\langle v, f(v) \rangle$ to the diagonal takes us to the point $\langle f(v), f(v) \rangle$; the small vertical arrow from that point to the curve takes us to $\langle f(v), f(f(v)) \rangle$; repeating the process—horizontal arrow to the diagonal, thence vertical arrow to the curve—takes us from the line "$y = f_n(v)$" to the line "$y = f_{n+1}(v)$". In order to discover the theorem we need to see that the limit of $f_n(v)$ as n increases is a fixed point of f, i.e. a z which equals $f(z)$. The picture *suggests* that the limit of $f_n(v)$ as n increases is the first such z greater than v. What (partly) visual thinking could constitute an epistemically acceptable route to this conclusion?

Let **v** be any element of the domain such that the curve at **v** is above the diagonal, i.e. so that $f(\mathbf{v}) > \mathbf{v}$. Let **z** be the first element of the domain greater than **v** that is a fixed point of f. We start with an argument by mathematical induction on the positive integers n that for all n, $f_n(\mathbf{v}) < \mathbf{z}$. Recall that the curve is increasing, so no two points on the curve are at the same height. Now suppose that $f_m(\mathbf{v}) < \mathbf{z}$; then the curve at the point when it has height $f_m(\mathbf{v})$ lies *above* the diagonal, otherwise the curve would have crossed the diagonal somewhere between $\mathbf{v}$ and $\mathbf{z}$, which contradicts the datum that $\mathbf{z}$ is the first fixed point of f greater than $\mathbf{v}$. Then there will be some (rightward) horizontal distance from the curve at that point to the point $\langle f_m(\mathbf{v}), f_m(\mathbf{v}) \rangle$ on the diagonal; then there will be some (upward) vertical distance between that point on the diagonal to the point $\langle f_m(\mathbf{v}), f_{m+1}(\mathbf{v}) \rangle$ on the curve, since the curve is increasing (i.e. as the nib tracing the curve moves rightwards, it moves upwards). Hence (i) if $f_m(\mathbf{v}) < \mathbf{z}$ then $f_m(\mathbf{v}) < f_{m+1}(\mathbf{v})$, as the second component of an ordered pair representing a point indicates its height above the horizontal axis. Also, if $f_m(\mathbf{v}) < \mathbf{z}$, the point $\langle f_m(\mathbf{v}), f_{m+1}(\mathbf{v}) \rangle$ lies on the curve to the left of point $\langle \mathbf{z}, f(\mathbf{z}) \rangle$, as the first component of an ordered pair representing a point indicates its (rightward) distance from the vertical axis; and so $f_{m+1}(\mathbf{v}) < f(\mathbf{z})$ as the curve is increasing; but $f(\mathbf{z}) = \mathbf{z}$; hence (ii) if $f_m(\mathbf{v}) < \mathbf{z}$, $f_{m+1}(\mathbf{v}) < \mathbf{z}$. From (ii) and the datum that $f_1(\mathbf{v}) < \mathbf{z}$, we can conclude that (iii) for all n, $f_n(\mathbf{v}) < \mathbf{z}$. See Figure A9.2 for illustration.

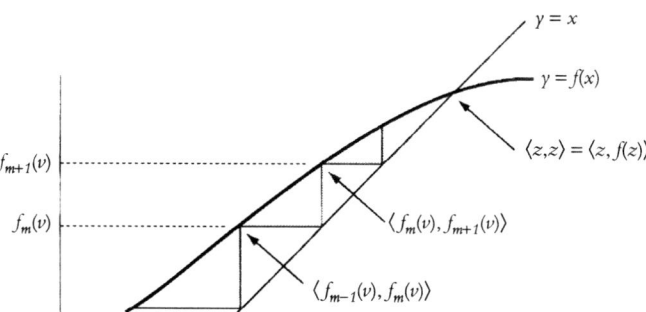

Figure A9.2

Now we need only show that for any positive ε there is an m such that for all $n > m$, $f(\mathbf{z}) - f_n(\mathbf{v}) < \varepsilon$. Take any positive real number ε. As f is continuous there is a positive number δ such that when $\mathbf{z} - w < \delta$, $f(\mathbf{z}) - f(w) < \varepsilon$. Now consider any $w < \mathbf{z}$ such that $f(\mathbf{z}) - f(w) \leq \varepsilon$. (Clearly there are such w; for example, $w = \mathbf{z} - \delta/2$.) If there is an m such that $f_m(\mathbf{v}) \geq f(w)$, it is clear that for all n greater than m, $f(w) < f_n(\mathbf{v}) < f(\mathbf{z})$, by (i) and (iii) above. Figure A9.3 illustrates. Hence for all n greater than m, $f(\mathbf{z}) - f_n(\mathbf{v}) < \varepsilon$, as desired. So it only remains to show that there is an m such that $f_m(\mathbf{v}) \geq f(w)$.

VISUAL THINKING IN BASIC ANALYSIS 189

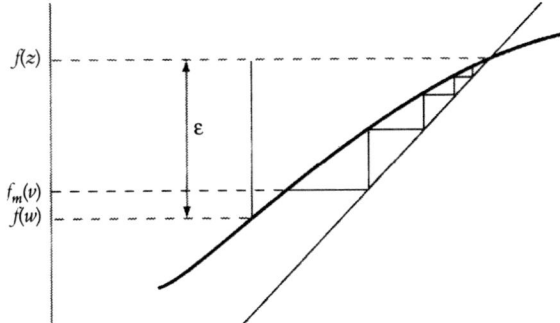

Figure A9.3

Since the curve over the closed interval $[\mathbf{v}, f(w)]$ does not touch the diagonal, there is a positive lower bound α on the lengths of vertical line segments between curve and diagonal over $[\mathbf{v}, f(w)]$. See Figure A9.4.

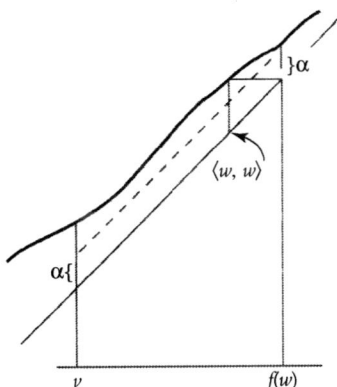

Figure A9.4

For any m the vertical distance $f_m(\mathbf{v}) - \mathbf{v}$ is the sum of the lengths of m vertical segments: $\mathbf{v}$ to $f(\mathbf{v})$, $f(\mathbf{v})$ to $f_2(\mathbf{v})$, ...$f_{m-1}(\mathbf{v})$ to $f_m(\mathbf{v})$. See Figure A9.5. If $f_m(\mathbf{v}) < f(w)$, each of these segments will be at least α in length, so $f_m(\mathbf{v}) - \mathbf{v} \geq m\alpha$. But if $f_m(\mathbf{v}) < f(w)$, it also follows that $f_m(\mathbf{v}) - \mathbf{v} < f(w) - \mathbf{v}$, hence that $m\alpha < f(w) - \mathbf{v}$. So if for every m, $f_m(\mathbf{v}) < f(w)$, it would follow that for every m, $m\alpha < f(w) - \mathbf{v}$, contradicting the Archimedean axiom. Hence for some m, $f_m(\mathbf{v}) \geq f(w)$, as needed.

This route to the theorem involves verbal and symbolic as well as visual thinking. But the visual thinking is not superfluous. At several points we

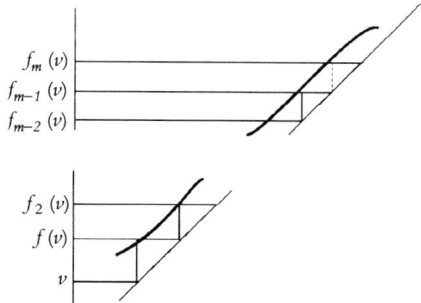

Figure A9.5

reach a claim by visual means, e.g. the claim (near the start) that the curve of f cannot remain above the diagonal over $[\mathbf{v}, 1]$ given that values of f are less than or equal to 1. So the non-visual parts cannot stand alone. But is the thinking epistemically acceptable, as it appears to be? I do not claim to have established that it is. It can only be so if the conditional assumptions linking visual with analytic conditions used from the start are known to be true by the thinker. Perhaps they can come to be known by sufficient mathematical experience of analysis, although I do not have a clear idea how this knowledge is delivered. My conclusion is that we cannot yet rule out the possibility that this route to the theorem is a route of genuine discovery. Even if it is, the theorem reached is much more restricted than those of Rolle and Bolzano. The bounds within which visual discovery in analysis is possible are set by the rare and highly restrictive assumptions linking the visual with the analytic that are known to be true by the thinker.

10
Symbol Manipulation

The aim of this chapter is to explore the nature and uses of visual thinking with symbols in mathematics. The question whether visual symbol manipulation can be a means of discovery will not be the sole focus of concern; this chapter looks also at how symbolic thinking contributes to other epistemological goals, such as security, illumination, and generality. The chapter starts with an account of the cognitive and epistemic nature of the kinds of symbolic thinking found in mathematics. To this end we need to draw some preliminary distinctions.

Preliminary distinctions

Concept-driven v. rule-driven symbol manipulation   Sometimes one's understanding of a symbol configuration, in particular one's grasp of the concepts expressed by operation symbols, enables one to recognize that certain changes of configuration preserve value. This occurs in arithmetic when it seems obvious that reordering and regrouping addends leaves the sum unchanged. Presumably the famous example of the 10-year-old Gauss was concept-driven. Given the task of finding the sum of the integers from 1 to 100, he transformed $1 + 2 + \ldots + 99 + 100$ into

$$(1 + 100) + (2 + 99) + \ldots + (49 + 52) + (50 + 51),$$

by imagining the numerals from 51 to 100 in reverse order beneath the numerals from 1 to 50.[1] I assume that his recognition that this transformation is value-preserving depended on an appreciation of the commutativity and associativity of addition that resulted from his concept of addition.

Now suppose one is given the task of finding 9/16 *divided by* 6/8. Many people's understanding of division does not lead easily to a way of calculating division by a fraction. Instead one may use a rule one believes reliable:

(R) If the divisor is non-zero, interchange its numerator and denominator, then multiply.

Solving the problem by following (R) is a case of rule-driven symbol manipulation, to be contrasted with concept-driven symbol manipulation. While concept-driven symbol manipulation is justified by one's understanding of the operation symbols, the acceptability of rule-driven symbol manipulation depends on one's justification for accepting the rule. In this regard there are three kinds of situation:

(1) One knows a justification for the rule and can produce the justification without difficulty. This is the situation of many people with respect to uses of rule (R).[2]

(2) One remembers having seen a justification of the rule but cannot recall or easily work out a justification. For some people an example would be the chain rule for differentiating a product of functions:

$gf'(x) = g'(f(x)) \times f'(x)$, where these factors exist.

(3) One takes the rule on trust and would have to get the justification from elsewhere, text or expert. An example might be use of Cramer's rule for solving regular systems of linear equations. We can rewrite the system (with unknowns x, y, and z)

$$a_1 x + b_1 y + c_1 z = d_1$$
$$a_2 x + b_2 y + c_2 z = d_2$$
$$a_3 x + b_3 y + c_3 z = d_3$$

as a vector equation with A, B, C, and D for the obvious column vectors:

$x\mathrm{A} + y\mathrm{B} + z\mathrm{C} = \mathrm{D}.$

And we can write |EFG| for the determinant of the matrix whose column vectors are E, F, G in that order. Provided that A, B, and C are independent, Cramer's rule allows us to put

$$x = \frac{|\mathrm{DBC}|}{|\mathrm{ABC}|}, \quad y = \frac{|\mathrm{ADC}|}{|\mathrm{ABC}|}, \quad z = \frac{|\mathrm{ABD}|}{|\mathrm{ABC}|}.$$

Arriving at a conclusion using Cramer's rule in circumstance (3) is like relying on the testimony of someone else: one would not have come to see the truth of the conclusion for oneself, at least not unconditionally. Even in circumstance (2) one would be relying on the testimony of a former self. Thus there are four grades of epistemic strength that an act of symbol manipulation may have. Concept-driven symbol manipulation is strongest, followed by rule-driven symbol manipulation when one can produce a justification for the rule; one is in a weaker position when aware of having had knowledge of a justification but cannot recall it, and weaker still when one takes the rule on trust from a teacher or texbook. But in all these cases symbol manipulation can yield new knowledge. Perhaps it is only the first case, concept-driven symbol manipulation, that can preserve a sense of obviousness, immediacy, and illumination. This is not to deny the value of rule-driven symbol manipulation. Symbolic rules for calculation should not be assessed as if they were intended to deliver understanding; they should be prized if they enable us to compute values with cognitive efficiency, that is, little effort, fair speed, and no errors.

Formal v. informal symbol manipulation The rule stated above for dividing by fractions is not a rule of purely *formal* symbol manipulation, since it is independent of the particular expression of the terms. That is, the rule can be applied to distinct expressions of the same division, and the actual symbol rearrangements required will differ. Compare for example the following transformations.

(a) $\dfrac{9}{16} \div \dfrac{3}{4}$ (a') $\dfrac{9}{16} \times \dfrac{4}{3}$

(b) $\dfrac{9/16}{3/4}$ (b') $9/16 \times 4/3$

Both (a) to (a') and (b) to (b') can be made by following (R). But a purely formal rule for the first will specify vertical interchange of the numerals on the right of the symbol "÷" and a replacement of that symbol by "×"; a formal rule for the second will not specify these things. Instead it will specify a horizontal interchange of the numerals beneath the symbol "—"; replacement of that symbol by "×" and horizontal alignment of all symbols, the complex numerical terms separated by the symbol "×" enclosed by single spaces; a formal rule for the first will not specify these things.

An advantage of rules of formal symbol manipulation is that they can be applied asemantically, that is, without any thought of what the composite symbol or its constituents express or denote. Thus one source of possible error is removed. This can be a significant advantage because making changes in a formal two-dimensional array of symbols is relatively easy for us, as is checking that such changes conform to given formal rules, provided that the changes are not too numerous and the symbol display is not too complicated. Of course, formal rules cannot be used to make new discoveries or prove old ones unless they can be justified, and their justification has to be semantic. But justifying formal rules can be a separate one-off episode.

Formal system and proof

In order to examine symbol manipulation in greater detail, I will introduce a formal system and an example of a formal derivation in that system. I will first specify a language for a fragment of vector algebra, taking as given a set **V** of vectors and a set **R** of scalars.[3] The language consists of a recursively defined set of terms, and equations. The *terms* in this system are variables, or expressions of the form "(σ @ τ)" where σ and τ are terms and "@" marks the place of a binary operator. (Note that "=" is not an operator but a relation symbol.) I will suppress outermost brackets around terms in what follows. A term is said to be *for* a class Γ when it is intended to designate a member of Γ or to range over all and only the members of Γ.

Terms and equations

> (a) "A", "B", "C", and all other upper case letters of the alphabet are variables for **V**.
> (b) "**r**", "**s**", "**t**", and all other bold lower case letters of the alphabet are variables for **R**.
> (c) If σ, τ are terms for **V**, an expression of the form "(σ + τ)" is a term for **V**.
> (d) If σ, τ are terms for **V**, an expression of the form "(σ · τ)" is a term for **R**.
> (e) If σ, τ are terms for **R**, an expression of the form "(σ + τ)" is a term for **R**.

An *equation* in this system is constituted by an "=" sign flanked by terms.

The theorem generator consists in a set of equations designated as axioms and rules for obtaining new terms from old.

Axioms and rules

> [1] $(\mathbf{r} + \mathbf{s}) + \mathbf{t} = \mathbf{r} + \mathbf{s} + \mathbf{t}$
> [2] $\mathbf{r} + (\mathbf{s} + \mathbf{t}) = \mathbf{r} + \mathbf{s} + \mathbf{t}$
> [3] $A \cdot B = B \cdot A$
> [4] $A \cdot (B + C) = (A \cdot B) + (A \cdot C)$

By substituting occurrences of a term τ for all occurrences of a variable *v* in one of the listed equations, one gets a *substitution instance* of that equation, provided that τ is for the class over which *v* ranges.[4] For simplicity I will allow that substitution instances of an axiom are also axioms. So the *axioms* of the system are [1] to [4] *and* their substitution instances.

The rules are for obtaining a term from a given term:

[R1] One term flanking an "=" sign in an axiom may be obtained from the other.

[R2] Let γ be a term occurring within another term σ; let us denote the term that results from replacing every occurrence of γ in σ by another term δ by "σ(γ/δ)". If γ and δ are terms flanking "=" in an axiom, σ(γ/δ) can be obtained from σ.[5]

Thus the term "(B · A) + (C · D)" may replace the term "(A · B) + (C · D)" using axiom [3]. Finally we need to specify the class of theorems.

Derivations and theorems

A *derivation* of a term β from a term α is a finite sequence of terms beginning with α and ending with β, each after the first being obtainable from its immediate predecessor by one of the rules. An equation α = β is a *theorem* if and only if it is an axiom or there is a derivation in the system of β from α or of α from β.

That is the formal system. Here is a slightly abbreviated sample derivation, one by which a certain theorem of vector algebra[6] can be proved. The axioms and rules used are indicated on the right.

(D + E) · (D + E)	
((D + E) · D) + ((D + E) · E)	by [4] and [R1]
(D · (D + E)) + (E · (D + E))	by [3] and [R2], *twice*
((D · D) + (D · E)) + ((E · D) + (E · E))	by [4] and [R2], *twice*
(D · D) + (D · E) + (E · D) + (E · E)	by [1] and [R1], *then* [2] *and* [R2]
(D · D) + (D · E) + (D · E) + (E · E)	by [3] and [R2].

These steps should seem straightforward. The first thing to notice is that we do not have to worry about the semantic values of expressions. We do not need to know which sets are denoted by "**V**" and "**R**"; we do not need to know the denotations of "D" and "E", nor do we need to know what operations are designated by "+", "+", and "·". Formal moves in reasoning are merely syntactic.

But formality is not all or nothing. The example departs from absolute formality in suppressing outermost brackets and presenting two steps as one. This is not a totally trivial matter. We tend to get lost in a plethora of brackets and we tend to get distracted when progress is very slow. This means that the security to be gained from formalization is constrained. The rules must be sufficiently flexible to minimize performance error resulting from high symbol density (e.g. too many brackets or subscripts) and from long sequences of very similar expressions. In practice a low degree of formalization is optimal most of the time.

The derivation allows us to take as a formal theorem of the system the equation:

$$(D + E) \cdot (D + E) = (D \cdot D) + (D \cdot E) + (D \cdot E) + (E \cdot E).$$

Reaching this equation in this way constitutes *proving* it only in a context in which the formal system is known to be sound. Soundness of a system is relative to a class of permitted interpretations:

> A system is *sound* if and only if every formal theorem of the system is true under every permitted interpretation of the language of the system.

To show that our system is sound it suffices to show that the axioms are true under every permitted interpretation, and that the rules preserve identity.[7] What counts as a permitted interpretation is a matter of stipulation: the relevant stipulations constitute the semantics of the system.[8]

When a formal system is invented to provide a regimented version of a body of informal mathematics, the semantic stipulations will be designed to recover our understanding of the informal mathematics. Why then bother with formalizing a body of mathematics? The advantages are twofold: one, to enable us to prove things about the body of mathematics, e.g. that it is consistent relative to another body of mathematics; two, in cases where our informal understanding is tenuous, to enable us to establish a particular theorem without relying on our understanding in the process of proof. But outside logic we are not interested in claims about a whole corpus of mathematics, as opposed to questions within one; and the cases in which we need to resort to formality to make a particular finding secure are quite rare. Moreover, there are considerable disadvantages in proceeding formally, namely, the extreme length and slowness of the process, and insecurity from the likelihood of slips when symbol arrays are large and complicated. For this reason almost all symbol manipulation in practice is informal.

In the next couple of sections, the derivation given in this section will be used to investigate the cognitive nature of particular kinds of symbol manipulation that occur in both formal and informal contexts.

Syntactic forms and substitution

At the level of formality used here it is fairly easy to see that the transitions can be made in accordance with the rules. Yet the thinking involved in making these transitions is quite complicated. To explain how we might do it I postulate a visual ability to extract certain syntactic forms common to distinct terms. A form of the sort I have in mind is a horizontal sequence of places for terms separated by operator expressions or brackets. For example, taking a letter string **vvv** or **xxx** or **zzz** to mark a place for a term, **vvv · xxx** is a form shared by the following terms:

A · (B + C),
A · B,
(B · (B + C)) · (B + C),
A · A.

I use strings of the same letter in a form-display to mark places for occurrences of the same term, so **xxx · xxx** is a form of "A · A" but not of "A · B". Strings of different letters are for occurrences of terms which *may* be different but do not have to be. A term may have more than one form, e.g., "A · (B + C)" has the forms

vvv · (xxx + zzz) and **vvv · xxx**.

Now let us return to the derivation given earlier. Consider the first step:

(D + E) · (D + E)
((D + E) · D) + ((D + E) · E).

How do we make this step? My hypothesis is that we find a form shared by the initial term and a term flanking "=" in one of the listed axioms. Then guided by common forms we make appropriate substitutions in the axiom, thereby obtaining a new axiom which licenses a move from the initial term to the next one by the rule [R1]. In this case we spot forms that the initial term shares with the term on the left in axiom [4],

$$A \cdot (B + C) = (A \cdot B) + (A \cdot C)$$

The left term has two forms in common with the initial term of the derivation,

vvv · (xxx + zzz) and **vvv · xxx**.

To use axiom [4] it is the first form that is relevant, so we must see that the initial term too has this form. This is easy for us.

$$\begin{array}{ccc} A \cdot (B & + & C) \\ \mathbf{vvv} \cdot (\mathbf{xxx} & + & \mathbf{zzz}) \\ (D+E) \cdot (D & + & E) \end{array}$$

In seeing "$(D + E) \cdot (D + E)$" as sharing this form with "$A \cdot (B + C)$", the left occurrence of "$(D + E)$" in the former is seen as occupying the same place in the form as "A" in the latter; hence we understand that "$(D + E)$" must be substituted for "A" throughout [4]. As the right occurrence of "D" is seen as occupying the same place in the form as "B", we understand that "D" must be substituted for "B" throughout; similarly "E" must be substituted for "C" throughout, giving the axiom

$$(D+E) \cdot (D+E) = ((D+E) \cdot D) + ((D+E) \cdot E).$$

Rule [R1] allows us to pass from the left-hand term to the right-hand term, as in the first step.

Now let us look at the last step:

$$(D \cdot D) + (D \cdot E) + (E \cdot D) + (E \cdot E)$$
$$(D \cdot D) + (D \cdot E) + (D \cdot E) + (E \cdot E). \qquad by\,[3]\ and\ [R2]$$

In this case we focus on a subterm of the initial term. Subterms are found by decomposing a term step by step, first obtaining terms by eliminating the main operators "+"; then removing brackets and eliminating the secondary operators "·". We select the third term (reading from left) in the first layer of components "$(E \cdot D)$" and see that it has the same form as the left-hand term in axiom [3]:

[3] $A \cdot B = B \cdot A$

The terms "A · B" and "E · D" are easily seen to share the form **vvv · xxx**. As "E" in "E · D" is seen as occupying the same place in the common form as "A" in "A · B", and "D" the same place as "B", we take it that "E" should be substituted for "A" and "D" for "B" throughout axiom [3] to get the following axiom:[9]

E · D = D · E

Given this, rule [R2] allows replacing the third term "E · D" in the penultimate term in the derivation by "D · E", thus obtaining the final term.

This account raises two questions about the cognitive nature of this kind of thought. How do we find common forms? Is the ability to see that a given term has a certain form purely visual, hence independent of linguistic abilities? There are several ways of finding common forms. The terms in the listed axioms, however, are intended to stand for particular forms, or at least to suggest them. More precisely, each term flanking an occurrence of "=" in a listed axiom is intended to suggest primarily the unique form that has each variable occupying a distinct place in the form, occurrences of the same variable occupying places indicated by the same letter string. Call this the *primary form* of a term. The distinction between primary forms and non-primary forms can be illustrated with a couple of examples.

Term	Primary form	Non-primary form
A · (B + C)	**vvv · (xxx + zzz)**	**vvv · xxx**
(A · B) + (A · C)	**(vvv · xxx) + (vvv · zzz)**	**vvv + xxx**

Hence one way in which common forms might be found is this. Keeping in mind the primary forms of right- and left-hand terms in listed axioms, we look to see which of those forms is a form of the term under consideration. In the case of the starting term of the derivation:

(D + E) · (D + E).

Almost immediate are the forms

vvv · xxx   and   **(vvv + xxx) · (vvv + xxx)**.

The first one is the primary form of each composite term in axiom [3]:

[3]   A · B = B · A

Applying [R1] merely gives us a self-identity, as it would amount swapping the two occurrences of "(D + E)", and so is unproductive. So one would look to see if the starting term has another form in common with one of the primary forms of listed axiom terms. Visual aspect shift is the key to this step. It is well known that presented with a rectangular array of black dots in columns and rows, if the intercolumn spaces are larger than the inter-row spaces, normally sighted people will tend to see the array as composed of columns.[10] But even if it is initially seen as composed of columns because of the larger intercolumn spacing, the viewer can override this at will to see it as an array of rows, provided that the ratio of intercolumn to inter-row spacing is not too great. In such cases one is exercising a voluntary capacity for visual aspect shift. We can exercise this same capacity in coming to see that the initial term has as one of its forms the primary form of the left-hand term of axiom [4]:

$$A \cdot (B + C) \quad vvv \cdot (xxx + zzz) \quad (D + E) \cdot (D + E)$$

Making the appropriate substitutions in axiom [4] and applying [R1] yields a new term, which we can take as the next term in a derivation.

Is the ability to see that a given term has a certain form purely visual, hence independent of linguistic abilities? At this point I am unaware of data that could decide the matter. There is a pertinent study[11] of syllogistic reasoning using forms (a) with content and (b) without content, such as

(a) All poodles are pets;   (b) All P are B;
    All pets have names;   All B are C;
    All poodles have names.   All P are C.

The task in all cases was to say whether the presented argument was deductively valid. The findings of that study indicated involvement of two dissociable neural networks depending on the presence or absence of content. During content-based reasoning a left hemisphere temporal lobe system was recruited (including a perisylvian language area used in semantic processing). This was not used in the no-content condition on structurally identical tasks. Instead a parietal system was used, as one would expect with visuo-spatial tasks. However, in both conditions a frontal area implicated in natural language syntax processing was involved. This, however, does not answer our question about forms of purely non-verbal terms. This is because evaluating the no-content syllogisms does require processing the syntactic role of natural language quantifier words such as "all", "every", "no", while no such words occur in the terms occurring in our fragment of algebra. As far as I know, the question is open and ripe for investigation.

Relocation, copying, insertion, and deletion

Let each term flanking the occurrence of "=" in an axiom be called a *main term* of the axiom. In the account just given, a term τ can be derived from a term σ by (a) extracting a form common to σ and a main term in a listed axiom, (b) using this form to make appropriate substitutions in the axiom to get a new axiom with σ as a main term, and (c) applying one of the rules in conjunction with the new axiom to obtain τ from σ. Sometimes, I believe, one does actually go through a procedure with steps (a), (b), and (c) in deriving a new term from a given term, especially in the early stages when one is unfamiliar with the system. But it is most improbable that one continues to go through the whole procedure each time. Instead, one uses rules permitting cognitive short cuts. These short-cut rules can be shown to be legitimate, in the sense that whenever a term τ can be obtained from a term σ by applying a short-cut rule, τ can be obtained from σ by means of the apparatus of the system without the use of short-cut rules.

Here is an example. We can treat axiom [3] as a rule that allows us, when given a term of the form **vvv · xxx**, to move the terms either side of the symbol "·" round so that they swap positions. In applying the rule, one might simply imagine a rotation in the plane of the dot-flanking terms that keeps them upright like seats in the big wheel at a fair, as indicated in Figure 10.1.

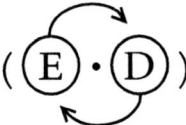

Figure 10.1

This short cut is legitimate, because any new term we can obtain this way from a given term we can obtain by making appropriate substitutions in axiom [3] and applying rule [R1]. Suppose a given term has a *constituent* term of the form **vvv · xxx**. Just rotate the dot-flanking terms around in the constituent to get a new term. This again is legitimate, because we can obtain the new term by making appropriate substitutions in axiom [3] and applying rule [R2]. Combining the two, we use a rule that allows us to rotate the dot-flanking terms in a term of the given form, whether or not that term occurs as a constituent of another term. We might operate this rule in reaching the last term from the penultimate:

$(D \cdot D) + (D \cdot E) + (\mathbf{E \cdot D}) + (E \cdot E)$
$(D \cdot D) + (D \cdot E) + (\mathbf{D \cdot E}) + (E \cdot E)$

Another example is provided by the steps from the third term in the derivation presented earlier to the fourth. The fourth term is obtained by two steps. One of these replaces the first of the following terms by the second:

$E \cdot (D + E)$
$(E \cdot D) + (E \cdot E).$

Again, in place of obtaining a new axiom from a listed axiom by making appropriate substitutions and then applying one of the rules [R1] and [R2], we can use a rule that allows us to transform the first term into the second. In this case we operate on a fragment of the first term that is not itself a term, namely, "E ·". We may imagine this splitting into two copies, which are then moved so as to become prefixed to the terms flanking the bold plus sign; then that bold plus sign is replaced by an ordinary plus sign. The first of these may be visualized as indicated in Figure 10.2:

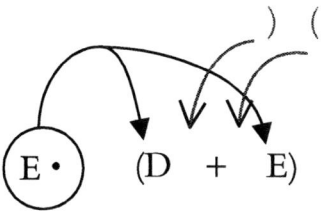

Figure 10.2

This example involves symbol-relocating movements, copying, and symbol insertions. Removing brackets is another kind of symbol manipulation, deletion. Deletion encompasses symbol manipulations prominent in algebra, e.g. those licensed by cancellation laws in group theory:

$b \cdot c = b \cdot d \Rightarrow c = d.$
$c \cdot b = d \cdot b \Rightarrow c = d.$

Leibniz's Law entails the converse of these laws and so permissible deletions are matched by permissible insertions. Again the inverse law of group theory

$$(b^{-1})^{-1} = b$$

licenses us to delete or insert "$(^{-1})^{-1}$" around a variable.

Substitution, relocation, copying, deletion, and insertion are among the major classes of symbol manipulation. Some or all of these are performed in visual imagination, when moving from one term or formula to another. It is likely that in *some* cases, especially symbol relocation, the visualizing has a motor element. Kosslyn distinguishes between motion-encoded and motion-added transformations in visual imaging.[12] If you see a galloping horse, you may later recall this in visual imagination. In that case the visualized motion of the horse is encoded. If on the other hand you recall seeing a static and solid coffee cup, you may imagine it gradually and continuously deforming into a doughnut. As the motion is supplied not by your memory but is a contribution of visual imagination following recall of the cup, that would be a motion-added transformation. In such a case you are not simply playing back a previously seen event, though your ability to add the motion in visual imagination may depend on your having seen motions of that type. The cases of visualized symbol movement described here are motion-added transformations.

But this classification itself divides into (a) cases in which one imagines that the movement is something that is brought about by oneself and (b) cases in which it is not imagined as brought about by oneself. It seems to me that when I visualize a coffee cup deforming I do not imagine this transformation as something that I am causing. But when I visualize a coffee cup rotating about a vertical axis, it seems I can imagine this as a motion initiated by me and even under my control. So among motion-added transformations some but not others are imagined as visualizer-caused. And among these there is a further division. In some cases the visualizer imagines the motion being caused by the action of one of his or her own body parts, so that the visual image includes the motion of the body part. An example of this would be visualizing the motion of a tennis racket in performing a particular stroke. In other cases one does not imagine causing the movement by moving a body part. The hypothesized classification is displayed in Figure 10.3.

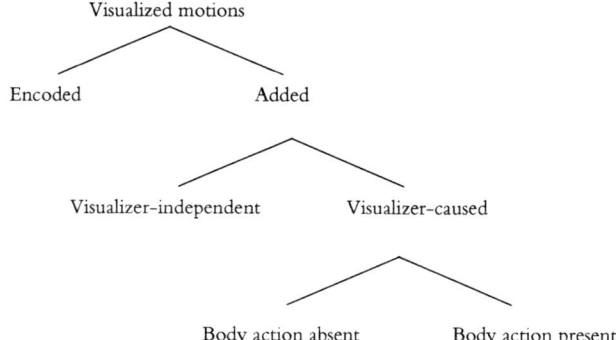

Figure 10.3

Where do the types of symbol manipulation discussed earlier belong in this classification? I conjecture that visualized substitution, relocation, copying, deletion, and insertion are imagined as visualizer-caused, probably without body action imaging. The very expression "symbol manipulation" suggests movements of symbols made by hand, and it is possible that some hand actions are imagined, even if this is not something we are aware of. If this were right we should expect to find a focus of neural activation during mathematical symbol manipulation in the posterior parietal lobe, where the dorsal pathway of visual processing that subserves action is found. I do not know of data that bear directly on this.

That is all that will be said here about the nature of symbol manipulation. What follows is an account of uses of symbol manipulation in mathematics. Use of symbol manipulation in determining the value of a composite term and in discovering or proving an algebraic equation has been lightly touched on already. Four other uses will now be discussed: getting a new goal, getting a strategy for a given goal, defining a kind of structure, and finding instances of a kind of structure.

Goal and strategy

A real number r can be represented in an n-dimensional vector space[13] as the rightward-pointing vector of length r along the x-axis. If A and B are such vectors, $(A + B)$ is just the vector that results from extending A along the x-axis by a segment of the same length as B. In this case there is an easy geometric way of reaching the theorem that

$$\|(A + B)\|^2 = \|A\|^2 + 2(A \times B) + \|B\|^2$$

where ||X|| is the length of X, illustrated in Figure 10.2:

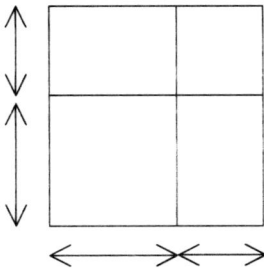

Figure 10.4

The question naturally arises whether this generalizes to

$$\|(A+B)\|^2 = \|A\|^2 + 2(A \cdot B) + \|B\|^2$$

for *any* vectors A and B, where ||X||, the length of X, is defined to be $(X \cdot X)^{1/2}$. This illustrates how use of symbols can give us a new goal.

From this definition it follows that $\|X\|^2 = (X \cdot X)$. So our question, re-expressed, is whether the following is true: For any vectors A, B

$$(A+B) \cdot (A+B) = (A \cdot A) + (A \cdot B) + (A \cdot B) + (B \cdot B).$$

Hence the goal is to establish this equation. A visual geometrical route to this equation is not at all obvious. Much easier is to see whether one of the familiar symbol-manipulating ways of establishing the theorem for real numbers can be mimicked to yield the vector generalization. This is the strategy. So here is a pair of important advantages of symbolic thinking in mathematics. First, it often gives us new goals: to find whether a theorem expressed by a certain formula generalizes to one which can be expressed by the same formula or one very similar to it, when the domain is enlarged and the operations appropriately extended. Secondly, it provides us with an initial strategy in these cases: see how far one can get towards establishing the generalization by following the sequence of symbol-manipulation moves which led to the original theorem.

Defining structure

A well-known advantage of formal or semi-formal symbolic thinking is that it facilitates the study of abstract structure. Here is an example. The following are familiar truths about real number addition and multiplication.

Associativity: $x + (y + z) = (x + y) + z$
$x \cdot (y \cdot z) = (x \cdot y) \cdot z$

Commutativity: $x + y = y + x$
$x \cdot y = y \cdot x$

Distributivity: $x \cdot (y + z) = (x \cdot y) + (x \cdot z)$

Existence of identity elements:[14]
> There are v and w such that for any x, $x + v = x$ and $x \cdot w = x$

Existence of inverses:
> For every x there is a y such that $x + y = v$, for v as above.
>
> For every x but v there is a y such that $x \cdot y = w$, for v, w as above.

These are structural facts about the real numbers with respect to addition and multiplication. It is sometimes thought that the laws of associativity and commutativity are merely optional conventions of notation. If that were so, we could adopt the negation of each of these laws for real number addition and multiplication without risk of falsehood. To see that this is not possible, think what a counter-instance to the commutativity of addition would entail. There could be two line segments of differing lengths such that the total length of the segment obtained by appending one to the other depends on which end was taken as the zero end, the starting point for measurement.[15] If it were merely a matter of optional conventions we could adopt laws of associativity and commutativity for real number subtraction and division without risk of falsehood, but again, it clearly is not so.

Now we can consider *any* domain **D** closed under *any* binary operations $\oplus$ and $\otimes$ for which exactly corresponding laws hold. The formal symbolism makes it easy to say precisely what the "corresponding" laws would be. They are the propositions expressed by the formulas given above when the variables are taken to range over domain **D** and the operation symbols "+" and "·" are taken to stand for the operations $\oplus$ and $\otimes$ respectively. In this way we use the formulas to define a structural kind. A domain **D** with binary operations $\oplus$ and $\otimes$ that has this structure is known as a *field*.

Many structural kinds are specifiable in this manner, e.g. groups, rings, lattices; and we can put specifications together in order to define further kinds of structure. For example if a set **V** forms an abelian group with respect to a binary operation **+**, and **D** is a field with respect to binary operations $\oplus$ and $\otimes$, (**V**, **+**) is a *linear space* over (**D**, $\oplus$, $\otimes$) just when for any a in **D** and any x in **V** ax is an element of **V** and, for any a, b in **D** and any x, y in **V**,

$$a(x+y) = ax+ay$$
$$(a \oplus b)x = ax+bx$$
$$a(bx) = (a \otimes b)x$$
$$1x = x, \text{ where 1 is the } \otimes \text{-identity element of } \mathbf{D}.$$

A small digression: too many distinct operation symbols make this difficult to absorb. We can avoid this by using disjoint sets of letter variables for **V** and for **D**. Then we can use "**+**" for both **+** and $\oplus$, and juxtaposition of variables for $\otimes$ as well as for multiplication of an element of **V** by an element of **D**, as in "*ax*". This does not invite confusion because we can tell which operation is signified by the type of variables. Then the first three formulas are visually simpler and easier to remember:

$$a(x+y) = ax + ay$$
$$(a+b)x = ax + bx$$
$$a(bx) = (ab)x$$

Absolute formality requires that distinct operations are denoted by distinct symbols. So here is further evidence that symbolic thinking is cognitively optimal when it is not totally formal.

This definition of linear spaces is just one illustration of a use of symbolic thinking that well proved its worth in twentieth-century mathematics. Formulas of a symbolic language are used to specify this or that kind of structure, and rule-governed transitions from term to term or from one formula to another are used to show that all structures of the specified kind have this or that structural feature. But it is wrong to think that one's grasp of a certain kind of structure consists entirely in knowledge of its formal definition. Grasp of a kind of structure is strengthened mostly through our awareness of particular instances—the greater the variety of instances, the stronger one's grasp. The possibility and nature of that kind of awareness is a large subject that will be investigated in the next chapter.

Grasping a kind of structure has a pragmatic as well as recognitional side. A strong grasp of what a Boolean algebra is, for example, entails both a facility to recognize that a given structured domain is or is not a

Boolean algebra and a facility to discover or at least to comprehend features of Boolean algebras by moving in thought between different instances; some of these will have characteristic visual representations which are not symbolic, e.g. circuit diagrams for electrical circuits, Venn diagrams for subsets of a point set. Our ability to discern structure and structural features is not primarily a symbolic-linguistic ability; it is cognition by capacities for analogy and generalization which are not yet understood. But defining a kind of structure is a symbolic-linguistic task that can be accomplished by means of a set of formulas within the context of a system of rules governing syntax (constraints on how the basic symbols can be arranged to form complex symbols) and semantics (constraints on how the formulas can be interpreted as statements), as illustrated above.

From means of study to objects of study

Arrays of symbols and manipulations of them are among the major *means* of mathematical thought. Symbol arrays can also lead us to *objects* of mathematical thought, as in matrix algebra. Given a system of linear equations for simultaneous solution, we can simplify the presentation by separating the coefficients from the unknowns. For example the following system of three linear equations in three unknowns

$$x + 5y + 2z = 9$$
$$x + y + 7z = 6$$
$$-3y + 4z = -2$$

becomes the following matrix equation

$$\begin{bmatrix} 1 & 5 & 2 \\ 1 & 1 & 7 \\ 0 & -3 & 4 \end{bmatrix} \begin{bmatrix} x \\ y \\ z \end{bmatrix} = \begin{bmatrix} 9 \\ 6 \\ -2 \end{bmatrix}$$

Ease of visual intake is not the main advantage. The 3 × 3 matrix invites one to switch attention from the coefficients within each equation in turn, i.e. the rows, to the coefficients for each unknown in turn, i.e. the columns. An important property of a matrix involves its columns: a matrix is *regular* if and only if it is of form $n \times n$ and its column vectors are linearly independent.[16] One way in which regularity is important is its significance for systems of linear equations: every system of linear equations with a regular matrix has a unique solution.

Given a matrix A and a vector v, if the number of A's columns equals the number of v's components, the operation of applying A to v is defined. So, for example, if v is the column vector (x, y) and A is the 3×2 matrix whose columns are first $(8, 0, 7)$ and second $(3, 5, 7)$, Av is the column vector $(8x + 3y, 0x + 5y, 7x - 7y)$.

$$\begin{bmatrix} 8 & 3 \\ 0 & 5 \\ 7 & -7 \end{bmatrix} \begin{bmatrix} x \\ y \end{bmatrix} = \begin{bmatrix} 8x + 3y \\ 5y \\ 7x - 7y \end{bmatrix}$$

One way of visualizing the operation is by moving and rotating v into a horizontal position, splicing it first into A's first row and interpolating a "+", then the same with subsequent rows in order, leaving adjustments (e.g. changing "$7x + -7y$" to "$7x - 7y$") to be made when writing out the result. Given a sequence of numerical values for the components of Av, say $w = (16 \cdot 5, 7 \cdot 5, 0)$, we get the matrix equation $Av = w$:

$$\begin{bmatrix} 8 & 3 \\ 0 & 5 \\ 7 & -7 \end{bmatrix} \begin{bmatrix} x \\ y \end{bmatrix} = \begin{bmatrix} 16 \cdot 5 \\ 7 \cdot 5 \\ 0 \end{bmatrix}$$

From these two matrix equations we recover the following system of linear equations:

$$8x + 3y = 16 \cdot 5$$
$$5y = 7 \cdot 5$$
$$7x - 7y = 0.$$

We can also think of the equation $Av = w$ as saying that the matrix A maps the 2-vector v onto the 3-vector w. Thus A itself represents a transformation from a two-dimensional vector space into a three-dimensional vector space. In general any $m \times n$ matrix[17] M represents a linear transformation from an n-dimensional vector space into an m-dimensional vector space.[18]

We can add $m \times n$ matrices by adding corresponding components to get another $m \times n$ matrix. And we can multiply an $m \times n$ matrix by a scalar c by multiplying the components by c to get another $m \times n$ matrix. These matrix operations interact with application of matrices to vectors in the following way:

$$(A + B)v = Av + Bv \quad \text{and} \quad (cA)v = c(Av).$$

Thus matrices interact with vectors in a rule-governed way which places them at the heart of linear algebra. But the important point for our purposes is that matrices are structured mathematical objects in their own right, rather than just ways of displaying numbers. We can define an $m \times n$ matrix by any way of assigning to each ordered pair (j, k) the element that in the standard matrix display is denoted by the symbol in the jth row and k^{th} column. Of course it is via instances of the standard matrix display that we become acquainted with matrices, just as we become acquainted with integers greater than 10 via arabic numerals in decimal form. In both cases the object, matrix or integer, is distinct from the expression, matrix table, or multi-digit numeral.

The components of matrices need not be numbers; they could for example be functions. In fact there is no restriction on the kind of domain from which matrix components can be drawn, as long as the chosen domain is closed under given "addition" and "multiplication" operations. Once the domain is chosen, $m \times n$ matrices themselves can be added, as noted earlier, and any $m \times n$ matrix A can be multiplied by any $n \times p$ matrix B to get an $m \times p$ matrix AB.[19] Matrix multiplication is associative and distributes over matrix addition; matrix addition distributes over matrix multiplication. For each n there is a two-sided identity element I_n for all $n \times n$ matrices, and every regular matrix has a unique inverse.[20] So the matrices with components drawn from a given domain constitute a space with its own algebra. We can of course study matrix algebra in the abstract without regard to the nature of matrix components. Thus the matrix display format has led us to a world of mathematical objects, matrices themselves, and their algebraic relations.

Conclusion

Symbol manipulation has several kinds of use. Rule-governed symbol manipulations can be used to compute values or to transform expressions into value-equivalent expressions. With a sufficient degree of formality the steps in a procedure (a computation or transformation) may be purely syntactic. Given a semantic justification of the rules used, performance of the procedure can deliver knowledge. The benefit of symbol manipulation procedures is cognitive efficiency: relatively high reliability, relatively small time and effort. The explanation of these benefits is that the procedures consist of clear sequences of what, for humans with normal visual abilities, are easy steps, such as extracting common forms, symbol insertions, relocations, deletions, and rotations—though some steps may involve complicated cognitive processing. But these benefits accrue only within a limited bandwidth of formality: too low and it may be unclear whether a particular move is correct; too high and the attentional cost of keeping track of changes in a crowd of symbols will be prohibitive.

Procedural efficiency is not the only benefit of symbolic thinking in mathematics. It can facilitate advances in the following way. We can investigate whether a formula proven to hold in one domain also holds in another domain; if the proof of the formula in the original domain can be set out as a sequence of symbol manipulations we can investigate whether that same sequence is valid when interpreted in the new domain, and if not, what changes are required. Thus we gain new goals and strategies. Symbolic thinking also provides generalizing power. A formal symbolic expression of principles governing given operations and relations of a particular domain, such as the arithmetic of real numbers, enables us to characterize a general kind of structure, such as fields. In the same way we can define a general kind of transformation, such as linear transformations of one vector space into another. Associated with this generalizing power are benefits of discovery. Provided we are careful to maintain consistency we can combine definitions of kinds to define new kinds, and as mentioned before, we can investigate the possibility that a formal analogue of a theorem about structures of one kind holds in structures of another kind. Finally, symbol displays and operations of symbol manipulation on those displays can reveal to us a new kind of object and new operations on objects of the new kind, as illustrated by the story of matrix algebra.

Notes

1. Hall (1970: 4).
2. One justification consists in the equivalence of the following equations for non-zero variables, together with an understanding of why each is equivalent to the next.

$$x/y \div v/w = z$$
$$x/y = z \times v/w$$

$$x/y \times w/v = z \times v/w \times w/v$$
$$x/y \times w/v = z.$$

3. The formal system is not sufficient for vector algebra. But vector algebra is not our concern. The purpose of the example is to illustrate points about formal systems and symbol manipulation.
4. e.g., the equation "$(B+C) \cdot D = D \cdot (B+C)$" is a substitution instance of axiom [3].
5. Rule [R2] is a syntactic rule corresponding to a restriction of Leibniz's Law. Otherwise known as the indiscernibility of identicals, Leibniz's Law is that if a is the very same thing as b, everything true of a is true of b and everything true of b is true of a.
6. $\|(A+B)\|^2 = \|A\|^2 + 2(A \cdot B) + \|B\|^2$, where $\|X\| =_{df} (X \cdot X)^{1/2}$, i.e. the length of X.
7. Here I am assuming that "=" stands for the identity relation in all permitted interpretations. A rule R *preserves identity* if and only if for any permitted interpretation I and any terms of the system τ and σ, τ is obtainable from σ by R (given the listed axioms) only if the denotation of τ under I is the very same as the denotation of σ under I.
8. I will not go into this here as it takes us too far from the concerns of this chapter. But any textbook in mathematical logic will give examples of semantics for a given formal system.
9. Recall that we have stipulated that substitution instances of axioms are axioms.
10. See Bruce et al. (1996: 106, 107).
11. Goel et al. (2000).
12. Kosslyn (1994: 350–3).
13. A vector space **V** is a linear space (V, +) over a field (D, +, ×) where V is the set of vectors and D is the set of scalars. Very often the scalars we use are real numbers.
14. Uniqueness of identity elements is easily proved. If v_1 and v_2 are identity elements with respect to +, $v_1 = v_1 + v_2 = v_2 + v_1 = v_2$. Similarly for "·" in place of "+".
15. I am talking about length of line segments in a Euclidean plane.
16. A vector *v* depends linearly on vectors $u_1, u_2, ..., u_n$ if and only if there are scalars (usually real numbers) c_i such that $v = c_1 u_1 + c_2 u_2 + ... + c_n u_n$.

A set of vectors is *linearly independent* if and only if none of them depends linearly on the rest.

17. The convention is: an $m \times n$ matrix has m rows and n columns.
18. Moreover, for any n- and m-spaces **V** and **W**, any linear transformation T:**V** → **W** is represented by an $m \times n$ matrix. For fixed bases of **V** and **W** the representation is unique. An $m \times n$ matrix A is a *linear transformation* if and only if for any, any n-vectors u, v, and any scalar c, $A(u + v) = Au + Av$ and $A(cv) = c(Av)$.
19. Letting b_i be the ith column of B, AB is the matrix whose ith column is Ab_i (where $1 \leq i \leq p$). An n-vector b can be taken to be an $n \times 1$ matrix, so application of an $m \times n$ matrix A to b can be thought of as a special case of matrix multiplication.
20. If a matrix A is $m \times n$, $I_m A = A I_n = A$.

11
Cognition of Structure

The realm of mathematics contains geometric figures and symmetry-preserving transformations; it contains numbers and arithmetical operations. It also contains groups of symmetries and rings of numbers. These latter things are examples of *structured sets*, that is, sets considered under specific relations, functions, or constants. A familiar example of a structured set is the set of natural numbers with functions of successor, addition, and multiplication, and the constant 0. For simplicity I will usually consider structured sets under a single binary relation. Given any set S of things, the set of subsets of S under the inclusion relation ($\subseteq$) is one example; a non-mathematical example is the set of English expressions under the relation "x is a syntactic constituent of y". In these cases the relation imposes on the set a structure, which is just the arrangement of the members of the set induced by the relation.

The structure of a structured set is independent of the nature of things related and independent of the nature of the specified relation: quite different sets under quite different relations have the same structure if the members of one set can be correlated one-one with the members of the other in a way that preserves the pattern of relations. A very simple example is this: (a) the set whose members are just the empty set, its unit set, and the set of those two things $\{\emptyset, \{\emptyset\}, \{\emptyset, \{\emptyset\}\}\}$ under the strict subset relation "$\subset$"; (b) the set $\{0, 1, 2\}$ under "$<$". The order preserving correlation is indicated in Figure 11.1 by the dashed arrows. Here is a definition:

> For a set A under binary relation R and a set B under binary relation S, a mapping **c** is an *order preserving correlation* $\equiv$ **c** is a 1-1 function from A onto B such that for any members x, y of A, x bears R to y if and only if $\mathbf{c}(x)$ bears S to $\mathbf{c}(y)$.

COGNITION OF STRUCTURE 215

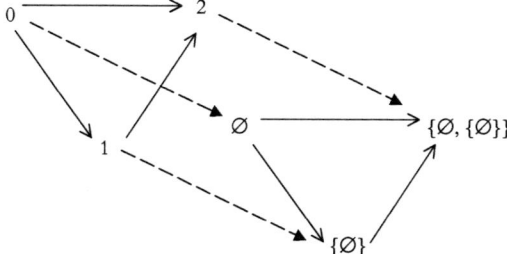

Figure 11.1

This definition can easily be generalized to cater for structured sets under several relations, functions, etc. When there is an order-preserving correlation between two structured sets, they are said to be *isomorphic*, which is a short way of saying that they have the same structure.[1] I will use the word "structure" to refer to the property that isomorphic structured sets have in common.

In this chapter I explore the nature of our cognitive grasp of structures. Some things we know by experiencing them or from information that derives ultimately from someone's experiencing them, such as our knowledge of persons. Knowledge by experience also includes knowledge of the *general types* of things exemplified by concrete objects or particular events, such as the shape of butterfly wings and the sound of thunder. In contrast to knowledge by experience there is knowledge via theory. Some things we have knowledge of only because they are posits of a theory we know to be true. Our knowledge of an unobserved astronomical body that is postulated to explain perturbations in planetary orbits is like that, knowledge via theory. In the latter case we know that exactly one thing satisfies some condition, and we know that thing as the entity that fulfils the condition, e.g. as the astronomical body that causes such-and-such perturbations in planetary orbits.

A plausible view is that we can know structures only via theory. Suppose we know that there is an interpretation of some set Σ of formal sentences in terms of a structured set under which all the sentences are true. In such a case the structured set is said to be a *model* of Σ. (For example, let Σ be a formal version of the second order Dedekind-Peano axioms for arithmetic; then the set of natural numbers under the successor function and the ordinary interpretations of "+", "×", and "0" is a model of Σ.) Suppose we also know that any two models of Σ are isomorphic (as we do for the Σ just mentioned). Then we know that exactly one structure is instantiated by models of Σ, and we can know this structure as "the structure common to all models of Σ". Thus we know the structure via a theory, and what we know about it is restricted to what we can deduce from its description as the structure of models of Σ (plus background knowledge). In many cases our knowledge of a structure is like that: we will know it as *the structure of models of theory X*. Structures are highly abstract; so it is reasonable to expect that this is the only way of knowing a structure. I will try to show that, on the contrary, this is not the only way of knowing structures; in some cases we can know structures in a more intimate way, by means of our visual capacities.

We can say, broadly, that our ability to discern the structure of simple structured sets is a power of abstraction; but we do not at present have an adequate account of the cognitive faculties that are operative in structure discernment. That we have an ability to discern structure is made evident by our manifest ability to spot a structural analogy. A striking case is provided by the history of biology. The pattern of *gene distribution* in dihybrid reproduction postulated by Mendelian theory can be found in the *behaviour of chromosomes* in meiosis (division of cells into four daughter cells, each with half the number of chromosomes of the parent) and subsequent fusion of pairs of daughter cells from different parents. The hypothesis that chromosomes are agents of inherited characteristics issued from this observation, and was later confirmed.[2]

Somehow biologists spotted the common pattern. But how? For an answer one might look to the large cognitive science literature on analogy. This subject is still in an early stage, with lots of speculation and many theories. Among the most promising are those based on the structure-mapping idea of Dedre Gentner.[3] This, however, is no good for present purposes, because it takes cognition of structure as given, and cognition of structure is precisely what we want to understand.

I will proceed by suggesting some possibilities consistent with what I know of current cognitive science, in particular vision science. I will restrict consideration to just a few structures, mostly very simple finite structures. But I will also say something about our grasp of infinite structures. Accordingly, the chapter is divided into two parts.

Visual grasp of simple finite structures

Here is another biological example of a structured set, simpler than those just mentioned. Consider the set consisting of a certain cell and two generations of cells formed from that initial cell by mitosis, division of cells into two daughter cells, under the relation "x is a parent of y". In this structured set there is a unique initial cell (initial, in that no cell in the set is a parent of it), a first generation of two daughters and a second generation of four granddaughters, which are terminal (as they are not parents of any cell in the set). How can we have knowledge of the structure of this structured set?

Visual templates

One possibility, highlighted by Michael Resnik, is that we use a perceptible template.[4] This might be a particular configuration of marks on a page, an instance of a diagram, which itself can be construed as a structured set, one that has the very same structure as the cell set. Let Figure 11.2 be our diagram.

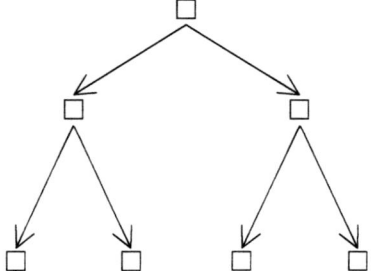

Figure 11.2

The structured set here is the set of square nodes of Figure 11.2 under the relation "there is an arrow from x to y". (One has to be careful here. We get a different structured set if the chosen relation is "there is an arrow *or chain of arrows* from x to y".) Now we let the top square node represent the initial cell, the square nodes at the tips of arrows from the top represent the daughter cells, and the four terminal nodes represent the granddaughter cells, with no two nodes representing the same cell. We can tell that these conventions provide an isomorphism:

> cell c is a parent of cell d if and only if there is an arrow from the node representing c to the node representing d.

218 COGNITION OF STRUCTURE

In this case we are using a *particular visible structured set* to represent another structured set. In the same way we can use the visible structured set to represent *all* structured sets that share its structure. This representative can thus serve as a visual template for the structure. We can see the visual template, and we can see it *as* a structured set. Seeing it as a structured set requires a non-visual factor, to determine what is the relevant set (e.g. square nodes or arrows) and what is the relevant relation (e.g. "there is an arrow from x to y" or "there is chain of one or more arrows from x to y"). But the result is still a kind of visual cognition.

Suppose now we are interested in some other structured set (like the cells under the *parent of* relation). By giving names to its elements and labelling the nodes of the visual template with these names, we can establish that the given structured set is isomorphic to the visual template. The labelled configuration is something we see, and checking that the labelling gives a one-one order-preserving correlation is a mechanical-perceptual task. From the finding that there is a one-one order-preserving correlation, an isomorphism, we safely infer that the structured set has the same structure as the visible template. In this way we have epistemic access to a structure.

Here I have been supposing that a visual template is one particular configuration of actual physical marks, e.g. those in Figure 11.2 in the copy of the book you are now reading! Just as you can recognize different physical inscriptions as instances of an upper case letter A, so you can recognize different configurations of physical marks as instances of the same *type*. You would have no difficulty in recognizing, as an instance of that type, the instance of Figure 11.2 in another copy of the book. Moreover, you could recognize as instances of that type other configurations that are geometrically similar but differing in size and orientation, as in Figure 11.3.

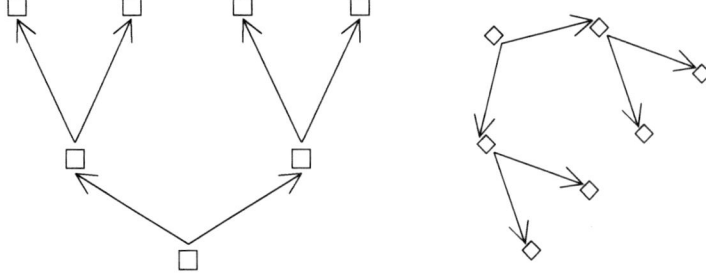

Figure 11.3

To account for this recognitional capacity it is thought that the visual system stores representations of *types* of visible configurations. A representation of a visual type is not itself a visual image; but

(i) a visual experience can activate a representation of a type—this is required for recognition as opposed to mere perception.
(ii) Also, we can activate a type representation at will to produce an image of an instance of the type—this is active visual imagining.

In this situation it seems reasonable to take the configuration *type* to have the structure-fixing role that we had been assigning to a particular physical instance of the type. This is still within the ambit of the template idea. In place of a single physical template, however, we allow as templates any configuration of marks that we can recognize as instances of the type.

Beyond visual templates

We can have an awareness of structure that is more direct than the visual template mode, at least in such very simple cases as this, two generations of binary splitting. The awareness I have in mind is not tied to a particular configuration type. Although a particular configuration of marks, viewed as a set of elements visibly related in a particular way, may serve a person as an initial instance of a structure, one may later think of the structure without thinking of it as the structure of configurations of just that type. Once one has perceived a configuration of marks as a structured set, one can acquire the ability to perceive configurations of *other visual types* as structured in the same way. For example, one has no difficulty in seeing the configurations of Figure 11.4 as structured in the same way, even though they appear geometrically quite dissimilar.

It is not clear how we do this. In these cases there is a visualizable spatio-temporal transformation of one configuration into another that preserves number of members and relevant relations between them. (Relevant relations are those picked out in conceiving the configurations as structured sets.) So, seeing two dissimilar configurations, conceived as structured sets, as instances of the same structure *may* involve visualizing an appropriate spatial transformation.

But I doubt that it is necessary to visualize a spatial transformation. We may instead be able to see, quite directly, diverse configurations as instances of the same structure. We can directly recognize objects of different shapes,

220   COGNITION OF STRUCTURE

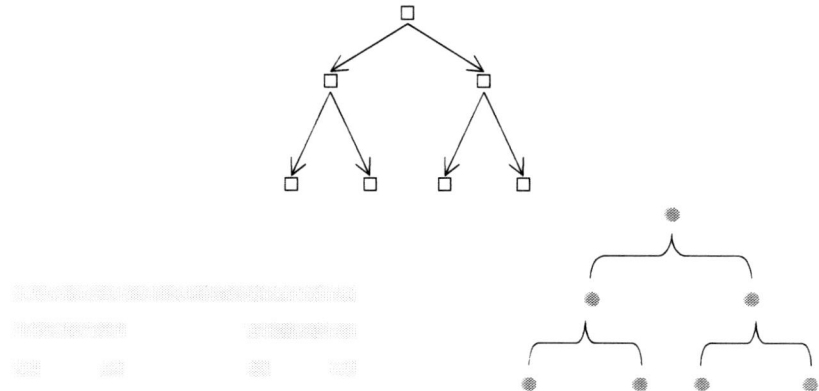

Figure 11.4

sizes, colours, surface textures, etc. as members of a single class, such as the category of *hand*. We sometimes think of all hands, or at least all left hands, as having the same shape. But the actual shape of a hand varies with positions of the fingers. What we actually mean by "the shape" of a hand is something more abstract, a spatial property that is preserved not only under shape preserving transformations, but also under transformations involved in normal changes of palm shape and finger positions. Yet we can visually recognize something as a hand directly, regardless of palm shape and finger positions. We don't have to *deduce* that something we see is a hand from other perceptible properties. The visual system has acquired a category representation for *hand*; and when visual inputs from seeing a hand activate this category representation, one visually recognizes what one is seeing *as* a hand. In the same way, the visual system can acquire a representation for a category of visual configurations of marks that provide instances of a common structure. In earlier chapters I have called these representations "category specifications". My suggestion is that a visual category specification gives us a visual means of grasping structure that is more flexible than a visual template.

A category specification is a collection of related feature specifications that cannot be "read off" the subjectively accessible features of the experience of seeing an instance of the category. Just as with face recognition, the subject may have no way of knowing precisely which congregations of features and relations lead to recognition of the category when presented with an instance. A visual category specification is a kind of visual representation, but it is very unlike a visual image or percept. An image or percept is a transient item of experience of specific phenomenological type, whereas a category specification is a relatively enduring representation that is not itself an item of experience. But activation of a category specification can affect visual experience, enabling visual recognition.

We can recognize a perceived configuration of marks as an instance of a certain structure, by activation of an appropriate visual category specification. Thus, I suggest, we can have a kind of visual grasp of structure that does not depend on the particular configuration we first used as a template for the structure. We may well have forgotten that configuration altogether. Once we have stored a visual category pattern for a structure, we have no need to remember any particular configuration as a means of fixing the structure in mind. We can know it without thinking of it as "the structure of this or that configuration". There is no need to make an association or a comparison. So this is more direct than grasp of structure via a visual template.

What about identifying the structure of a *non*-visual structured set given by verbal description? Recall the method mentioned earlier using a visual template: (1) we first name the members of the set, (2) then label elements of the template with those names, and (3) then check that the labelling provides an isomorphism. Though this *can* happen, we often do not need any naming and labelling. Consider the following structured set: {Mozart, his parents, his grandparents} under the child-to-parent relation "x is a child of y". Surely one can tell without naming and labelling that it is isomorphic to the structured sets given earlier. We know this for *any* set consisting of a person, her parents, and grandparents under the "child of" relation (assuming no incest). It is as if our grasp of these sets *as* structured sets already involves activation of the visual category specification for two generations of binary splitting.

Exactly the same applies to the set obtained from the first two stages in the construction of the Cantor set by excluding open middle thirds, starting from the closed unit interval, under set inclusion.[5] Naming and labelling are not needed to recognize this as a case of two generations of binary splitting. I offer as a tentative hypothesis that we can recognize the structure in this case as a result of activation of the visual category specification for two generations of binary splitting.

222 COGNITION OF STRUCTURE

Extending the approach: more complicated structures

The structured sets considered so far have all been very small finite sets under a *single binary relation*. You may reasonably harbour the suspicion that sets structured by a plurality of relations or operations lie beyond any visual means of cognizing structure. I will now try to allay that suspicion. Figure 11.5 is a labelled visual template for the structure of the power set of a three-membered set {a, b, c} under inclusion. The power set of a given set S is the set of all subsets of S; inclusion is the subset relation, denoted by "⊆". As before, it is a very small set structured by one binary relation.

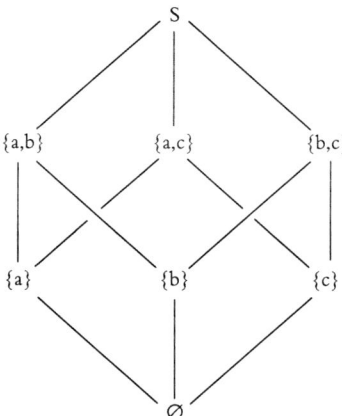

Figure 11.5

Let S be the set {a, b, c}; let ⟨**PS**; ⊆⟩ be the power set of S under inclusion. The configuration of Figure 11.5 itself provides a structured set: the set H of nodes under the relation "$n = m$, or, there is an upward path from m to n", which we can shorten to "$n \leq m$". This structured set is easily seen to be isomorphic to the power set of S under inclusion by means of the labelling in the figure. In symbols,

$$\langle H; \leq \rangle \cong \langle PS; \subseteq \rangle.$$

However, there is another way in which the set of nodes of Figure 11.5 is structured. In place of the binary relation ≤ there are operations and constants (or "distinguished elements") as follows:

(∧) $x \wedge y =$ the highest node n such that $n \leq x$ and $n \leq y$.
(∨) $x \vee y =$ the lowest node n such that $x \leq n$ and $y \leq n$.[6]

We name the bottom and top nodes, our constants, Min and Max respectively.

$(-) - x =$ the node n such that $x \wedge n$ is Min and $x \vee n$ is Max.

The set of nodes structured by $\wedge$, $\vee$, $-$, Max and Min is isomorphic to the power set algebra of S, which is the power set of S under the operations of intersection, union, and relative complement and the constants S and the empty set. In symbols

$$\langle H; \wedge, \vee, -, \text{Max}, \text{Min} \rangle \approx \langle \mathbf{P}\mathbf{S}; \cap, \cup, \sim, S, \emptyset \rangle.$$

The isomorphism can be checked visually.

The common structure of these structured sets is known as *the three-atom Boolean algebra*. Thus we have a visual template for the structure of a three-atom Boolean algebra, and this is a structure involving three operations, known as *meet* ($\wedge$), *join* ($\vee$), and *complement* ($-$), and two constants.

With a small amount of practice it is easy to acquire the visual ability to see the meet, join, and complement of nodes in Figure 11.5 right away. Perhaps the least straightforward is complement, but here is how it is done. Viewing the configuration as a cube, we see that every node is at one end of a unique diagonal of the cube; the complement of a node is the node at the other end of the diagonal. So, although it is an exaggeration to say that we can simply *see* the configuration of Figure 11.5 *as* the structured set $\langle H; \wedge, \vee, -, \text{Max}, \text{Min} \rangle$, it is strictly correct to say that we can have a perceptual grasp of that configuration under the relevant operations and constants. What now of our grasp of its structure, the three-atom Boolean algebra? Familiarity with instances can give us some awareness of the structure, but that awareness may not be sufficient for us to recognize instances when they are presented to us: we may need to map them onto a visual template such as the configuration of Figure 11.5.

Extending the approach: structural kinds

The cognitive abilities involved in discerning the structures of the example above can also be used to discern *kinds* of structure. Representing a structure by a spatial pattern can be useful when the structure is very small or very simple, but moderate size and complexity usually nullifies any gain. Try drawing a diagram of the power set algebra of a 4-element set: it can be done, but the result will not be easy to take in and memorize. Moreover, there is no easily acquired iterative procedure for obtaining a diagram for the power set algebra of an $n + 1$-element set from a diagram for the power set algebra of an n-element set. This means that the structures of the power set algebras of finite sets do not constitute a *kind* that we can grasp by means of visual configurations. Some kinds of structure, however, are sufficiently simple that visual representations together with rules for extending them can provide awareness of the structural kind. For discrete linear orderings with endpoints the representations might be horizontal configurations of stars and lines, giving a structured set of stars under the relation "x is left of y". The first few of these we can see or visualize, as illustrated in Figure 11.6.

Figure 11.6

We also have a uniform rule for obtaining from any given diagram the diagram representing the next discrete linear ordering with endpoints:

Add to the right end star a line segment with a star at its right end, maintaining horizontal alignment.

This is a visualizable operation that together with the diagrams of the initial cases provides a grasp of a species of visual templates for finite discrete linear orderings with endpoints. Through awareness of this kind of template we have knowledge of the kind of structure which they are templates for.

Binary trees, and more generally n-ary trees for finite n, comprise a more interesting kind of structure. We have already seen configurations that can be used to represent binary trees in Figures 11.2 and 11.3. Taking elements to be indicated by points at which a line splits into two, the first three binary trees can be visualized as in Figure 11.7.

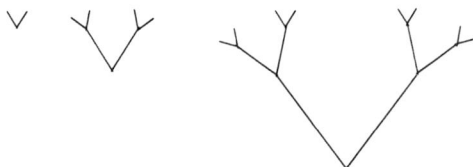

Figure 11.7

As mentioned before, a configuration represents a structured set only given some rules governing the representation of relations, functions, and constants. Binary trees are sets ordered by a single transitive relation R with one initial element and for every non-terminal element exactly two immediate R-successors.[7] I am restricting attention to binary trees such that, for some finite number k, all and only k^{th} generation elements are terminal.[8] The customary rule is this:

> x bears relation R to y if and only if there is an upward path from the node (branching point) representing x to the node representing y.

This convention clearly respects the transitivity of the relation.

The largest tree that we can easily visualize has very few generations. But we have a uniform rule for obtaining the next tree from any given n-generation tree by a visualizable operation:

> Add to each of the 2^n terminal points a "v".

This visualizable operation, together with the diagrams or images of the first few trees, provides us with a visual awareness of a kind of template for finite binary trees.

Whether or not we make the "v"s of each succeeding generation a fraction of the size of their predecessors (as in Figure 11.7), for some n which is not very large the parts of a tree diagram from the nth generation onwards will be beyond the scope or resolution of a visual percept or image in which the initial part of the tree is clearly represented. But we still have a visual way of thinking of a binary tree too large for its spatial representation to be completely visible at one scale. We can visualize first the result of scanning a tree diagram along any one of its branches, and we can visualize the result of zooming in on "v"s only just within visual resolution. This constitutes a kind of visual grasp of tree diagrams beyond those we perceive entirely and at once. This visual awareness, arising from the combination of our visual experience of the first few tree diagrams and our knowledge of the visualizable operation for extending them, gives us a grasp of the structural kind comprising the finite binary trees.

The same considerations apply with regard to ternary and perhaps even quaternary trees. But trees whose nodes split into as many as seven branches defy visualization beyond the first generation, and we cannot visualize even the first generation of a tree with twenty-nine branches from each node, except perhaps in a way in which it is indistinguishable from trees with thirty-fold branching. In these cases we operate by description and where possible by analogy with the trees we can visualize, just as we sometimes use three-dimensional objects as analogues of objects in higher dimensions. However, the sort of acquaintance we can have with the Binary-Tree kind we can also have with many other kinds of structure, if we can give a recursive specification of the kind. What is required is just that templates of the smallest two or three structures are easy to visualize and we have a rule for obtaining a template of the "next" structure from a template of a given structure by a visualizable process. A structural kind of this sort comprises an infinite sequence of structures that are often nested, in the sense that every structure in the sequence is a substructure of all later structures.[9] The union of such a sequence is an infinite structure. What kind of knowledge can we have of infinite structures? This question will be discussed next.

Knowledge of infinite structures

If there can be visual cognition of any infinite structure, the simple and familiar structure of the natural numbers will be the prime example. Here I am talking about the structure of the finite cardinals under their natural "less than" ordering. This structure, which I will call "$\mathcal{N}$", is shared by the set of arabic numerals of the decimal place system in their standard ordering,[10] the set of finite sequences of the letter "A" ordered by length, the set of von Neumann ordinals under the membership relation, and many others. It is easy to check that these are isomorphic by finding order-preserving correlations between them. Alternatively, one can know that they are isomorphic via an awareness that each of them can be mapped one to one in an order-preserving way onto a particular structured set with a visual representation. What I have in mind is a number line representation, as discussed in Chapter 6.

There are several possibilities for a number line. One is a set of evenly spaced vertical marks on a horizontal line, with a single leftmost mark continuing endlessly to the right such that every mark, however far to the right, is reachable by constant rate scanning from the leftmost mark. One mark is taken to precede another if it is to the left of it; with respect to this ordering the leftmost mark is the only initial mark and there is no terminal mark.[11]

Clearly we could not see or visualize more than an initial part of any such line. When it comes to actual images, something like Figure 11.8 will

Figure 11.8

be the best we can do. But as explained in discussing the nature of a mental number line in Chapter 6, if we distinguish between two kinds of visual representation, category specifications and images, the idea that the visual system has a representation for a line that extends infinitely in one or both directions holds no paradox. A *visual category specification* is a set of related feature descriptions stored more or less permanently; a *visual image* is a fleeting pattern of activity in a specialized visual buffer, produced by activation of a stored category specification. What is impossible is an infinitely extended visual image. But it is possible, and not at all puzzling, that a category specification specifies a line with no right end, one that continues rightward endlessly. Of course, a momentary image generated by activation of that category specification will represent only a finite portion of the line; but the specification that the line has no right end ensures that rightward imagistic scanning[12] will never produce an image of a right ended line.

My suggestion is that our grasp of this structured set, the well-ordered set of evenly spaced marks on an endless horizontal line, issues (or *can* issue) from a stored visual category specification. But this is not achieved by "reading off" the descriptions of the category specification, since we have no direct access to those descriptions. Rather, as a result of having the category specification, we have a number of dispositions which, taken together, indicate the nature of the structured set it represents. These are dispositions to answer to certain questions one way rather than another. For example:

Given any two marks, must one precede the other? *Yes.*
Do the intermark spaces vary in length? *No.*
Is the precedence of marks transitive? *Yes.*
Can any (non-initial) mark be reached from the initial mark by scanning to the right at a constant speed? *Yes.*

But some questions will have no answer:

Is the intermark length more than a centimetre?

These answers tell us something about the nature of the mental number line as determined by the features specified in the category specification.

The answers entail that no mark has infinitely many predecessors; as the marks form a strict linear ordering, this entails that they form a well-ordering. So we can say that the structure of the mental number line is that of a well-ordered set with a single initial element and no terminal element. This is the structure $\mathcal{N}$.[13] My proposal is twofold. First, in having a visual category specification for the mental number line we have a grasp of the structured set consisting of its number marks under the left-to-right order of precedence. Second, we have, or can have, knowledge of the structure $\mathcal{N}$ as the structure of this "number line" structured set.

Knowing the natural number structure in this way is different, less direct, from the kind of knowledge of finite structures discussed earlier. In this case we cannot experience an entire instance of the structure. So this knowledge of structure does not consist in an ability to recognize instances and to distinguish them from non-instances. It is knowledge by description, rather than knowledge by acquaintance. But it does have an experiential element that distinguishes it from knowledge by a description of the form *the structure of models of such-&-such axioms*.[14] To help us appreciate how significant this difference is, it is worth examining a contrasting case.

An infinite structure beyond visual grasp

A contrasting case is the structure of the set of real numbers in the closed unit interval [0, 1], under the "less than" relation. I will call this structure "$\mathcal{R}$". We do have visual ways of representing $\mathcal{R}$, but I claim that they do not give us knowledge of it. I will now try to substantiate this claim.

An obvious thought is that we can think of $\mathcal{R}$ as the structure of the set of points on a straight line segment with left and right endpoints, when each point corresponds to a unique distance from one end, and the order of points is determined by the corresponding distance. We can certainly visualize a finite horizontal line segment, taking its points to be the locations of intersection of the horizontal line segment with (potential) vertical line segments; and we can visually grasp what it is for one such location to lie to the left of another. Why does not this give us a visual grasp of the structure $\mathcal{R}$?

The reason concerns points. If the points on a line constitute a set with structure $\mathcal{R}$ they must be not merely too small to be seen by us, but absolutely invisible, having zero extension. Neither vision nor visualization gives us any acquaintance with even a single point of this kind, let alone uncountably many of them. In addition to a geometrical concept of extensionless points we have a perceptual concept of points as tiny dots. Perceptual points do have extension; some parts of a perceptual point on the line would be nearer the beginning of the line than others, so a perceptual point does not lie at exactly one distance from the beginning. For this reason no line of juxtaposed perceptual points could have the structure of [0, 1] under the "less than" ordering. So we must make do with geometrical points. But thinking of a line as composed of geometrical points leads to numerous paradoxes. For instance, the parts of a line either side of a given point would have to be both separated (as the given point lies between them) and touching (as there is no distance between the two parts, the given point being extensionless). A related puzzle is that a line segment must have positive extension; but as all its components have zero extension, the line segment must also have zero extension, as multiplication of zero by any number gives zero. Yet another puzzle: The unit interval has symmetric left and right halves of equal length; but symmetry is contradicted by the fact that just one of the two parts has two endpoints—either the left part has a right endpoint or the right part has a left endpoint, but not both, otherwise there would be two points with none intervening.

Setting aside these puzzles, thinking of an interval of real numbers as a set of points composing a line segment cannot reveal a crucial structural feature, Dedekind-continuity (completeness). This is the condition that, for every set of real numbers, if it is bounded above it has a least upper bound, and if it is bounded below it has a greatest lower bound.[15] So this visual way of thinking of an interval of real numbers is unrevealing about that feature which distinguishes it structurally from an interval of rational numbers. Hence we cannot know the structure by thinking of it visually in terms of the points in a line segment.

Perhaps there are other visual ways of thinking of an interval of real numbers. But none that I can think of fares any better. One possibility is the set of all branches of the infinite binary tree under the following relation of precedence:

x precedes y if and only if, when the branches x and y first diverge, x goes to the left and y goes to the right.

We can identify branches with infinite sequences of 0s and 1s. The left successor of each node is assigned 0, the right successor is assigned 1;

nothing is assigned to the initial node. The branch up to a given node is identified with the sequence of 0s and 1s assigned to nodes on that branch up to and including the given node; so, for example, the branch up to the leftmost fourth generation node (after the initial node) is ⟨0, 0, 0, 0⟩. A single infinite path up the tree, a branch, represents the infinite sequence of 0s and 1s assigned to its nodes. Each infinite sequence of 0s and 1s is the binary expansion of a real number in the closed unit interval. So we can use branches to represent real numbers in the unit interval. This correlation of sequences with real numbers is not injective: some pairs of sequences represent the same real number.[16] But we can easily rectify the matter by cutting out redundant branches. Two branches represent the same real number if and only if, when they first diverge, the one that goes to the left will go right ever after and the one that goes right will go left ever after. Let us call branches that represent the same real number *twins*; let S be the set of branches with all rightward twins cut out. Then there is a one-one correlation of S with [0, 1]: a branch is correlated with the real number of which it is the binary expansion. S ordered by "<", the relation of precedence among branches, is isomorphic to [0, 1] under "less than"; in other words it has the structure $\mathcal{R}$.

Why does *that* not give us a visual grasp of the structure $\mathcal{R}$? Unlike the ordered sets mentioned at the beginning of this chapter (two generations of binary splitting), whose members were represented by nodes of a finite binary tree, the members of S are infinite branches of an infinite binary tree. Since we can see or visualize only a finite portion of an infinite binary tree, not even one element of S is represented in any visual image of it. We see or visualize at most finite initial segments of branches, and each finite initial segment is common to infinitely many different branches. It is true that there are some branches for which we can have a visual category pattern: for instance the leftmost branch for unending zeros. But we can only have finitely many category patterns, and that leaves most branches unrepresented.

One might reply that we have a visual appreciation of how the infinite binary tree continues from one generation of nodes to the next, thus giving us a grasp of the structure $\mathcal{R}$ similar to our grasp of $\mathcal{N}$. But that is mistaken. The category specification for the infinite binary tree provides the basis for an awareness of the structure of the set of *nodes* under the relation "there is a path up from x to y"; but that is very different from the structure of the set of *branches* in S under the relation of precedence "<" given earlier, and only the latter is the structure $\mathcal{R}$.

Perhaps reflection on the visual representations used when we think of S ordered by "<" visually reveals that the ordering is linear and dense; but it is not clear to me how they would reveal that the ordering is Dedekind-continuous.[17] If it cannot, that is another reason why thinking visually of an infinite binary tree does not give us the capacities that would warrant a claim to knowledge of the structure. I suspect that a stumbling block to any visually anchored grasp of $\mathcal{R}$ is continuity. If this is right, our knowledge of its structure is theoretical rather than experiential: we know it as the structure of (a slight extension of) a model of the axioms for a complete ordered field.[18] This would constitute a clear contrast between the kind of knowledge we have of the structure $\mathcal{R}$ and the kind of knowledge we have of the structure $\mathcal{N}$.

Visual grasp of structures beyond $\mathcal{N}$?

I have argued that we can have a grasp of the structure $\mathcal{N}$ that is anchored to a visuo-spatial representation, and that this gives us a kind of awareness of the nature of $\mathcal{N}$ that is unavailable for $\mathcal{R}$. Is the kind of awareness that we have of $\mathcal{N}$ available for other infinite structures? Or is our grasp of $\mathcal{N}$ an isolated case? In a passage discussing the reach of what Hilbert called finitary mathematics, Gödel suggests by implication that there are other infinite structures knowable with the same kind of immediacy as our knowledge of $\mathcal{N}$. Gödel begins as follows:

> Due to the lack of a precise definition of either concrete or abstract evidence there exists, today, no rigorous proof for the insufficiency (even for the consistency proof of number theory) of finitary mathematics. However, this surprising fact has been made abundantly clear through the examination of induction up to ε_0 used in Gentzen's consistency proof of number theory.

He continues:

> The situation may be roughly described as follows: Recursion for ε_0 could be proved finitarily if the consistency of number theory could. On the other hand the validity of this recursion can certainly not be made *immediately* evident, as is possible, for example in the case of ω^2. That is to say, one cannot grasp at one glance the various structural possibilities which exist for decreasing sequences, and there exists, therefore, no *immediate* concrete knowledge of the termination of

232   COGNITION OF STRUCTURE

every such sequence. But furthermore such *concrete* knowledge (in Hilbert's sense) cannot be realized either by a stepwise transition from smaller to larger ordinal numbers, because the concretely evident steps, such as $\alpha \to \alpha^2$, are so small that they would have to be repeated ε_0 times in order to reach ε_0.[19]

The significant implications of this passage for present concerns are that the step from ω to ω^2 is "concretely evident"; that, as one can "grasp at one glance" the structural possibilities for decreasing sequences in ω^2, one can have "immediate concrete knowledge" that all such sequences terminate; hence that the validity of recursion (induction) for ω^2 can be made "immediately evident", whereas the same is not true for ε_0 in place of ω^2.

Is ω^2 really knowable in the implied way? Let us first step back. The ordinal ω under the membership relation (which in set theory does duty for the relation of ordinal precedence) has the structure $\mathcal{N}$; in fact this structured set (ω under "$\in$") is normally what set theory uses to represent the set of natural numbers under "$<$". So we can represent it in the same way, by a horizontal string of evenly spaced marks, with a leftmost mark, running off to the right endlessly. I will call such strings "ω-strings".

How do we make the step from the structure $\mathcal{N}$, exemplified by a single ω-string, to the structure of ω^2? Imagine a *vertical* sequence of horizontal ω-strings, starting from the top, left-aligned and evenly spaced, proceeding downward endlessly, so that if, visualizing, we were to scan this sequence downward, new ω-strings would come into view, still evenly spaced, however far we continued; and if we were to scan upward from any ω-string, at constant pace from each ω-string to its predecessor, we would arrive back at the topmost ω-string. This constitutes imagining a two-dimensional array of marks that has a top edge (the first ω-string) and a left edge (the column of first elements of ω-strings) but is infinite rightward and downward. I will call such an array an "ω-square"—this is not a description, as the array has no right or lower edge. The ordering is taken to be as for western script: for any two elements, if they are on the same line the leftmost element precedes, and if on different lines the element on the uppermost line precedes. Of course, we can have no image of more than a finite part of an ω-square; our representation of it is, as of an ω-string, a visual category specification. Thus we can have the kind of grasp of the structure of ω^2 that we have of the structure $\mathcal{N}$.

COGNITION OF STRUCTURE 233

What about Gödel's further claims about knowledge of ω^2? These are that, as one can grasp at a glance the various structural possibilities for decreasing sequences, one can have immediate concrete knowledge of the termination of every such sequence; hence that the validity of induction up to ω^2 can be made immediately evident.

What I think Gödel had in mind here is that, using the ω-square representation, there is a visuo-spatial way of telling that every decreasing sequence of members of ω^2 terminates. From this fact the validity of induction up to ω^2 quickly follows—I presume that Gödel took this as background knowledge for his readers. But how can we tell that every decreasing sequence of members of ω^2 terminates? Thinking of ω^2 in terms of an ω-square as described earlier, it is obvious that for any decreasing sequence, among rows containing members of that sequence, there will be an uppermost row; and that of the members of the sequence in that row there will be a leftmost member—call it α. Then, recalling the ordering of ω^2, it is clear that α is the least member of the sequence: so the decreasing sequence terminates. This way of acquiring the knowledge is relatively immediate, and is concrete in the sense that it is experiential. This *may* have been what Gödel had in mind, but we do not know. Either way, it does substantiate his claims.[20]

Thinking of the structure of ω^2 in terms of the ω-square can by its immediacy provide a route to discoveries that we are less likely to make without them. You may know how all positive rationals can be placed in an ω-square, and how Cantor used this fact to show in a visual way that we can reorder the rationals as an ω-sequence. Let the rational number terms p/q (for positive integers p and q) be ordered thus:

p/q precedes m/n if and only if $q < n$ or, $q = n$ and $p < m$.

We can tell that this structured set has the structure of ω^2 by mapping the positive rational number terms onto the ω-square by putting each term p/q in the qth row (from top) and the pth column (from left). See Figure 11.9. Cantor's enumeration is in the order indicated by the arrows, omitting any rational number term equal to one that has already appeared; thus we skip over "2/2" as the number this denotes has already occurred in the guise of "1/1". So the enumeration begins as follows:

(1) 1/1, (2) 2/1, (3) 1/2, (4) 1/3, (5) 3/1, (6) 4/1, ...

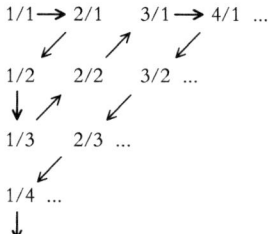

Figure 11.9

Clearly this is a one-one correlation of the positive rationals with the positive integers. This entails that there are no more rational numbers than integers, a surprising fact given that between any two rational numbers there are infinitely many others.

This example shows that we have an experiential way of telling of a structured set that it has the structure of ω^2. What about non-instances? It is not difficult to tell that instances of the structure $\mathcal{N}$, for example, are not instances of the structure of ω^2. An instance of $\mathcal{N}$, such as ω, has just one initial element, while every instance of the structure of ω^2 has infinitely many initial elements, represented by the leftmost elements in each row of an ω-square. So any one-to-one correlation of ω with ω^2 must correlate some initial elements of ω^2 with non-initial elements of ω. This entails that the correlation cannot preserve order. Figure 11.10 is an attempt to show why not, by showing how any correlation of $\omega \times 2$ (the first two ω-strings of ω^2) with ω disrupts order: the crossing of two correlation lines indicates that for some x and y, x precedes y in $\omega \times 2$ but the correlate of x does *not*

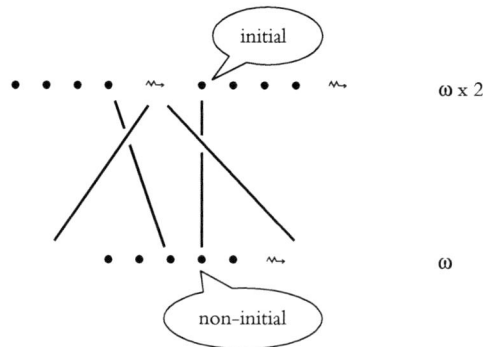

Figure 11.10

precede the correlate of γ in ω. Each squiggly arrow represents an infinite sequence of dots following the dots to its left.

Investigation is needed to determine how much further into the transfinite we may go before this kind of acquaintance with ordinal structure becomes impossible. At present I do not see how to extend the kind of account I have suggested for ω^2 to ω^ω. But I think that we can form a visual category specification for ω-many layers of ω-squares, an ω-cube, thus forming a representation of ω^3. There would be a top ω-square, and beneath each ω-square another one. The ordering within ω-squares is unchanged; for elements α and β in different ω-squares, α precedes β if and only if α's ω-square is above β's ω-square.

This uses our natural representation of space as extending infinitely in each of three dimensions. As our natural represention of space lacks a fourth dimension, it is clear that we cannot get to ω^4 from ω^3 in the same way as we got to ω^3 from ω^2. What we can do is to take each element of an ω-cube to represent an ω-string, an ω-square, or an ω-cube, thus getting a way of thinking of ω^4, ω^5, or ω^6. But this is a way of *using* a visual category specification for ω-cubes, not a way of extending it to get another visual category specification. Can we imagine putting an ω-cube in each position in an ω-cube,[21] so as to get a visual representation of ω^6? Not if this is visuo-spatial imagination, as opposed to supposition. This is because a representation of an ω-cube is a representation of something with infinite spatial extension, while each element of an ω-cube is represented as finitely extended. If, however, "imagine" is understood broadly to include fictional supposition, we can imagine each element of an ω-cube to be a benign black hole into which we can dive; and once in there, we find ourselves in a new infinite three-dimensional space containing another ω-cube (and of course one could iterate the story). Although this does not amount to forming an integrated visuo-spatial representation of an ω-cube of ω-cubes, it does give us a semi-visual way of thinking of higher powers of ω. But even this semi-visual thinking runs out before ε_0.

There may be other kinds of infinite structures, non-ordinal structures, that are knowable with the kind of visual awareness that we have for the structure of ω and ω^2 under their standard ordering. $\mathcal{N}$, the structure of ω, is the cognitively simplest infinite structure. We have usable visual representations of instances of it that provide some awareness of the nature of the structure. But these representations are category specifications, not

236   COGNITION OF STRUCTURE

images. The fact that we can never experience an image of the whole of an instance of this structure marks a qualitative difference between our grasp of this simplest infinite structure and our grasp of the finite structures discussed earlier, such as the two-generation binary tree or the power set algebra of a three-element set.

Conclusion

The central claim of this chapter is that we can have cognitive grasp of some structures by means of visual representations. For small simple finite structures we can know them through visual experience of instances of them, somewhat as we can know the butterfly shape from seeing butterflies. But most structures, even most finite structures, are too big and too complicated for visual cognition. Such a structure can be known only in a theoretical way, typically as the structure of all models of some particular theory. Surprisingly though, some infinite structures can be known by visual means. Of course, we cannot see or visualize more than a finite fragment of any instance of an infinite structure, but a sufficiently simple infinite structure may be represented by a visual category specification. The images produced by activation of a category specification and our dispositions to answer questions about what it represents may enable us to determine the structure and to discern isomorphisms between its instances. Thus a kind of visual cognition is available even for some infinite structures.

Notes

1. The details in full generality can be found in textbooks of model theory, abstract algebra, and mathematical logic.
2. Priority is attributed to W. Sutton (1902).
3. Gentner (1983).
4. Resnik (1997: ch. 11 §2).
5. This is the set whose members are: the closed interval of real numbers [0, 1]; the two closed outer thirds of [0, 1], namely [0, 1/3] and [2/3, 1]; and the four closed outer thirds of those two, [0, 1/9], [2/9, 1/3], [2/3, 7/9], [8/9, 1]. A set S *includes* a set T if and only if T is a subset of S.

6. The operation $x \wedge y$ is sometimes known as the *meet* or the *infimum* of x and y; the operation $x \vee y$ is sometimes known as the *join* or *supremum* of x and y.
7. So the ordered sets represented by the configurations in Figures 11.2 and 11.3 are not binary trees: the ordering relations were "x is a parent of y" among cells and "x is a child of y" among humans, and neither of these is transitive. But those same sets ordered by "x is a forebear of y" and "x is a descendent of y" respectively are binary trees.
8. The initial element is zeroth generation, its two successors are first generation, and so on. Accordingly, a tree whose terminal elements are first generation is a 1-generation tree, and in general a tree whose terminal elements are n^{th} generation is an n-generation tree.
9. ⟨A, R⟩ is a *substructure* of ⟨B, S⟩ if and only if A ⊆ B and for all x and y in A, if x bears R to y, x bears S to y.
10. What is meant here is the system without a decimal point, starting with "0" and proceeding lexicographically excluding any sequence of two or more arabic digits beginning with a "0"; the standard ordering is this (almost) lexicographic ordering.
11. An *initial* element is one that has no immediate predecessor; a *terminal* element is one that has no immediate successor.
12. Imagistic scanning is not inspecting in sequence the parts of a fixed image, but continuously changing the image in a way that is subjectively like scanning. The image change results from change of parameter value inputs during activation of the category specification. See Chapter 6 for a fuller explanation.
13. Mathematical logicians are understandably preoccupied by the fact that the set of *first-order* Dedekind-Peano axioms for number theory has models not well ordered by their "<" ordering; in those models there are sets isomorphic to the integers under "<", whose elements follow all the finite elements. If we want to rule out these non-standard models axiomatically we replace the first-order induction axiom with the original second-order axiom. Non-standard models are ruled out by the category specification as it dictates that any non-initial mark can be reached from the initial mark by scanning to the right at a constant speed, and that intermark spaces do not vary. Of course certain second-order assumptions are built in to the underlying conceptions of space, time, and motion here; but these are our natural conceptions.

14. Such knowledge requires knowing that the axiom set is categorical, meaning that all its models are isomorphic.
15. To say that a set B of real numbers is *bounded above* means that there is some real r such that for all b in B, $b < r$. Such an r is said to be an *upper bound* of B. The *least upper bound* of B is that upper bound of B that $<$-precedes all other upper bounds of B. The definitions of lower bound and greatest lower bound are similar.
16. Consider the sequence starting with 0 and continuing with 1s thereafter. This can be loosely thought of as representing the "infinite sum"

    $$1/4 + 1/8 + 1/16 + \ldots + 1/2^n + \ldots$$

    This is really the limit of the series of the finite sums

    $$1/4, 1/4 + 1/8, 1/4 + 1/8 + 1/16, \ldots, \sum_{1 < k \leq n} 1/2^k, \ldots$$

    which is exactly 1/2. But this is also the limit of the series

    $$1/2, 1/2 + 0, 1/2 + 0 + 0, \ldots, 1/2 + n \times 0, \ldots$$

    which has a binary expansion beginning with 1 and continuing thereafter with 0. So the two sequences, 0 followed by 1 recurring and 1 followed by 0 recurring, are correlated with the same real number. In general two sequences are correlated with the same real number if and only if they are identical up to and including their n^{th} components but the remainder of one of them is 0 followed by 1 recurring while the remainder of the other is 1 followed by 0 recurring.
17. A relation "$x < y$" on a set S is a *linear ordering* if and only if it is an irreflexive transitive ordering such that for any x, y in S $x = y$ or $x < y$ or $y < x$. A linear ordering "$<$" on S is *dense* if and only if for any x, z in S there is a y in S such that $x < y$ and $y < z$. The "less than" ordering of the set of rational numbers is a dense linear ordering that is not Dedekind-continuous; the same holds for any interval of rational numbers. The "less than" ordering of the set of real numbers is a Dedekind-continuous ordering; the same holds for any interval of real numbers.
18. "Complete" here is just another term for Dedekind-continuity. The axioms for a complete ordered field are given in Apostol (1967: vol. i, introduction, pt 3). For proof that all models of the axioms are isomorphic, see Cohen and Ehrlich (1963). The slight extension is

required because the closed unit interval has first and last points. The interval without those endpoints has the structure of the real numbers (under "<"), so we only need to add a first and last position to get the structure of $\mathcal{R}$.

19. Gödel (1972). The italics are Gödel's own. ω is the least infinite ordinal. (An ordinal is a set well ordered by the membership relation "$\in$" such that all its members are subsets of it.) $\langle \omega; \in \rangle$ has the structure $\mathcal{N}$. ω^2 is the first ordinal after $\omega \times n$ for every finite n, and $\omega \times n$ is the ordinal consisting of a sequence of n ω-sequences. Let α be any ordinal and define $\alpha \uparrow n$ thus: $\alpha \uparrow 0$ is α; $\alpha \uparrow n+1$ is $\alpha^{(\alpha \uparrow n)}$. Then ε_0 is the first ordinal after every $\omega \uparrow n$ for finite n. By "recursion for ε_0" Gödel meant induction up to ε_0: If for every $\alpha < \varepsilon_0$, Φ is true of α if Φ is true of every $\beta < \alpha$, then Φ is true of every $\alpha < \varepsilon_0$.

20. Robert Black pointed out in discussion that it does not support Gödel's implied claim that one can grasp at one glance all the structural possibilities for decreasing sequences in ω^2. For there are decreasing sequences with elements that occur arbitrarily far to the right in an ω-square. Gödel's other claims, including his main claim about knowledge of induction up to ω^2, are untouched by this.

21. This suggestion was made in discussion by Stewart Shapiro.

12

Mathematical Thinking: Algebraic v. Geometric?

In this final chapter I want to step back in order to get a perspective on thinking in mathematics as a whole. An examination of mathematical thinking of any serious depth and scope reveals a striking diversity. Is there any natural and illuminating way of classifying the kinds of thinking found in mathematics? How should the prospective taxonomist of mathematical thinking approach the task? We commonly draw a distinction between *algebraic* thinking and *geometric* thinking. While that classification may suffice for casual discussion in restricted contexts, I claim that the algebraic-geometric contrast, so far from being a dichotomy, represents something more like a spectrum. To the extent that there is a fundamental dichotomy in mathematical thinking, it appears to be between spatial and non-spatial thinking. But I will present reasons for dissatisfaction with any binary classification: we should aim instead to develop a much more discriminating taxonomy of kinds of mathematical thinking based on detailed cognitive research.

Algebraic versus *geometric thinking?*

One problem with the algebraic-geometric distinction is its obscurity. We do not have an operational criterion for deciding in which of the two categories this or that example of mathematical thinking belongs; instead we go by subjective impressions. Here is the kind of example that prompts us to contrast geometrical with algebraic thinking, one of the many arguments for Pythagoras's Theorem. It begins with Figure 12.1.

From this figure it is easy to see that the area of the larger square is both $(a+b)^2$ and $2ab + c^2$. Thus one reaches "geometrically" the equation

$$(a+b)^2 = 2ab + c^2.$$

ALGEBRAIC V. GEOMETRIC THINKING   241

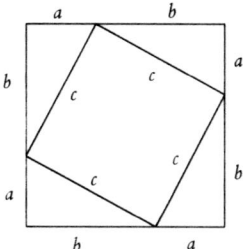

Figure 12.1

From here one proceeds "algebraically" as follows:

$a^2 + 2ab + b^2 = 2ab + c^2.$
$a^2 + b^2 = c^2.$

Then we turn our attention back to the figure, and take a final "geometrical" step: the square of the hypotenuse of a right-angled triangle equals the sum of the squares of its other two sides. With this example in mind we can try to characterize the geometric-algebraic distinction. It does not seem to be a question of subject matter, as that does not seem to change with the algebraic steps. This is just as well if the distinction is to apply across the board, in areas as diverse as number theory and topology, as is intended. But now, if we press for clarification, trouble appears. A common response is that algebraic thinking is essentially *symbolic*, while geometric thinking is essentially *spatial*. The problem with this characterization is that it fails to carve out exclusive kinds. Symbolic thinking typical of algebra, to wit, rule-governed manipulation of symbols, is just as spatial as geometrical thinking. The rearrangements, additions, and deletions of symbols discussed in Chapter 10 and exemplified in this little argument for Pythagoras's Theorem are operations in a two-dimensional space. Moreover, these operations depend on spatial features of the input symbol array and may be performed independently of any semantic content assigned to the symbols.

Symbolic and spatial thinking, then, are not mutually exclusive; on the contrary symbolic thinking falls within spatial thinking. Within the scope of spatial thinking, however, there does seem be a contrast between *thinking with diagrams* and *thinking with symbols*, i.e. formulas, terms, and their constituents. Perhaps this is what people have in mind by the algebraic-geometric distinction in mathematical thought. But is there a sharp dividing line between diagrammatic and symbolic thinking? Let us look at some examples.

1. *Left-cancellation law in Group Theory* Here is a proof of this law:

$$a \cdot x = a \cdot y$$
$$a^{-1} \cdot (a \cdot x) = a^{-1} \cdot (a \cdot y)$$
$$(a^{-1} \cdot a) \cdot x = (a^{-1} \cdot a) \cdot y$$
$$e \cdot x = e \cdot y$$
$$x = y.$$

Our concern is with the cognitive operations in proceeding from one line to the next. Consider, for example, the relocation of brackets involved in the step from the second line to the third. Presumably this is a paradigm case of symbol manipulation. Syntactically specifiable moves take us from one sentence to a new sentence.

2. *Matrix multiplication*

$$\begin{pmatrix} a & b & c \\ d & e & f \end{pmatrix} \begin{pmatrix} x \\ y \\ z \end{pmatrix} = \begin{pmatrix} ax + by + cz \\ dx + ey + fz \end{pmatrix}$$

The thinking I have in mind here is the visualizing of an operation on the left side configuration to get the right-hand matrix. One way of doing this is to visualize the column vector moving and rotating into a horizontal position, splicing it first into the upper row of the left matrix and interpolating "+" signs to get the upper term of the right matrix, then doing the same to the lower row of the left matrix to get the lower term of the right matrix.

This, like the moves in establishing left cancellation in Group Theory, is clearly a case of symbol manipulation. Yet the symbol arrays for the matrices are not ordinary terms. Each one *displays* the structure of the matrix it stands for. That is, spatial relations between the letters of the matrix array represent relations between the elements of the matrix, so that these relations are simultaneously displayed rather than sequentially stated. Thus a two-dimensional matrix array has a diagrammatic element. But the whole equation is not a diagram. It is rather more like a sentence written with an ideographic script.

ALGEBRAIC V. GEOMETRIC THINKING 243

3. *Grafting trees of associativity*[1] Figures 12.2 and 12.3 present some configurations used in a branch of algebra with associated geometrical objects known as associahedra (or Stasheff polytopes). One can get from one tree to the other by visualizing rotation out of the page around a page-vertical axis through a vertex, or by visualizing the short inner segment slide down the main V and up the other side.

Figure 12.2

Figure 12.3 shows two of the five possible 4-branch trees. Again one can visualize transforming one into the other by a single movement, a rotation or movement of a segment across to the other side of the main V.

Figure 12.3

Operations of grafting allow us to obtain new trees from given trees. Figure 12.4 gives an example.

Figure 12.4

The operation γ_4 grafts the second tree onto the fourth branch of the first. In general γ_i grafts a j-branch tree onto the ith branch of a k-branch tree to obtain a tree with $k+j-1$ branches. Each tree is a two-dimensional object, and the spatial arrangement of the branches displays a particular structure of association. But the trees can still be treated as terms in an equation and the syntactic character of the moves is obvious.

4. *Using commutative arrow diagrams* This example shows how thinking with commutative arrow diagrams can lead to a simple theorem. An arrow diagram is *commutative* just when any two paths along arrows or

244 ALGEBRAIC V. GEOMETRIC THINKING

chains of arrows with the same starting point and the same endpoint represent equal arrows. Thus commutative arrow diagrams carry equational information. Here are definitions of two expressions I will use in stating the theorem.

(1) An arrow r is a *retraction* for f just when r·f is the identity arrow on the domain of f.
(2) An arrow s is a *section* for f just when f·s is the identity arrow on the co-domain of f.

In Figure 12.5 commutative arrow diagrams represent these situations. The diagram on the left says that s is section for f. The diagram on the right says that r is a retraction for f. The lower diagram in Figure 12.5 is an amalgamation of the upper two; it says that the arrow f has both a retraction r and a section s.

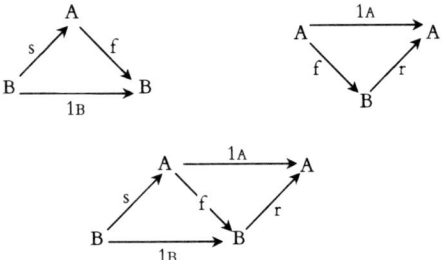

Figure 12.5

Here is the theorem:

If f has both a retraction r and a section s, r = s.

By the laws of identity arrows, the path from B to A along s followed by 1_A represents an arrow equal to s, and the path from B to A along 1_B followed by r represents an arrow equal to r. In short

$1_A \cdot s = s$
$r \cdot 1_B = r$.

These facts can represented by adding arrows to the lower diagram of Figure 12.5 so as to get Figure 12.6. As any two arrows or paths of arrows in a commutative diagram that start from the same object and end at the same object represent equal arrows, s = r. Thus we have arrived at

ALGEBRAIC V. GEOMETRIC THINKING   245

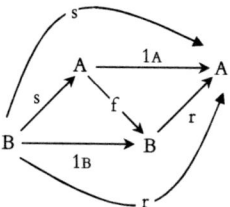

Figure 12.6

the theorem. The theorem is trivial, but familiarity with commutative arrow diagrams makes it almost immediate: visualizing the amalgam will suffice.

Arrow diagrams—I omit "commutative" for brevity—are further removed from the linguistic mould than matrix multiplication. An arrow diagram is significantly unlike a linguistic presentation in two ways:

- Relationships are simultaneously displayed rather than sequentially stated.
- Spatial relations between parts of the diagram indicate relations between the things represented by the parts. Thus the arrow diagram of Figure 12.6 is a unified representation of a complex situation.

The same information in sentence form would require several atomic sentences, and we would have to piece together the bits in order to get a unified grasp of the situation. This is a typical difference between sentential descriptions and diagrams. But there are also properties in common:

- Arrow diagrams have syntax. They have terms (arrows, chains of arrows, letters for domains and co-domains), and certain arrangements of terms express equality.
- Discrete bits of information are carried by discrete parts of the diagram. For example, the fragment of Figure 12.6 reproduced in Figure 12.7 says that $r \cdot f \cdot s = r$.

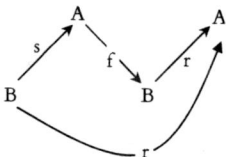

Figure 12.7

246 ALGEBRAIC V. GEOMETRIC THINKING

This is a property typical of collections of mathematical formulas, but not typical of diagrams. In these two respects arrow diagrams are like symbolic sentential constructions.

Using graphs of real-valued functions of reals   Recall that the derivative of one of the two functions cos(x) and sin(x) is just the other one, and the derivative of the other is *minus* the first one. If you cannot remember which is which, here is an easy way to find out. First sketch the graphs of the two functions from a little before 0 to π as in Figure 12.8.[2]

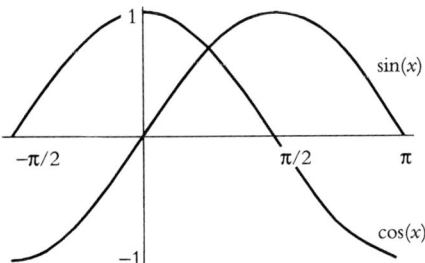

Figure 12.8

Note that cos(0) = 1. So, if the derivative of sin(x) is cos(x), the graph of sin(x) when x = 0 will have gradient 1, that is, it will have the slope of the ascending diagonal (y = x). But if the derivative of sin(x) is − cos(x), the graph of sin(x) at x = 0 will have gradient −1, that is, it will have the slope of the descending diagonal (y = −x). So we only have to look at the graph of sin(x) around 0: it has the slope of the ascending diagonal, not the descending diagonal; so the derivative of sin(x) is not −cos(x).

From our prior information, we can conclude that the derivative of sin(x) is cos(x), hence that the derivative of cos(x) is − sin(x). We can check this using the same method but swapping the functions. The sine wave has value 1 at around x = π/2. The cosine wave around x = π/2 coincides with the descending diagonal, as it should if the derivative of cos(x) = − sin(x).

The double graph has several features that set it apart from the representations of earlier examples. Here are some of them.

- Discrete bits of information are *not* carried by discrete parts of the graphs.

- Though some information is carried locally, e.g. that the graphs cross at about $\pi/4$, other information is carried globally, e.g. that $\cos(x)$ is about $\sin(x + \pi/2)$.
- Some information is merely approximate, as indicated by the word "about" in the previous point. In the example we use the fact that the gradient of $\sin(x)$ at 0 is roughly that of the ascending diagonal $y = x$ but not even roughly that of the descending diagonal $y = -x$.
- The double graph carries not only bits of information but also suggestions. For example, it suggests (but does not say) that there is a reflection symmetry around the vertical $x = \pi/4$, i.e. that $\cos(\pi/4 + x) = \sin(\pi/4 - x)$.

What we can infer about the functions from the information carried by the diagram depends on background information. For example, given that the functions are continuous, we can infer from the graph crossing that there is some value of x for which $\sin(x) = \cos(x)$; without that background datum we cannot infer this.[3] The four features just listed are typically lacking in symbolic representations in mathematics.

6. *Using geometrical depictions*   Let t be one of the tangents to the sphere passing through P, as in Figure 12.9. Problem: Is the distance along t between its point of contact with the sphere and P the same as the distance along any other tangent through P between *its* point of contact with the sphere and P?

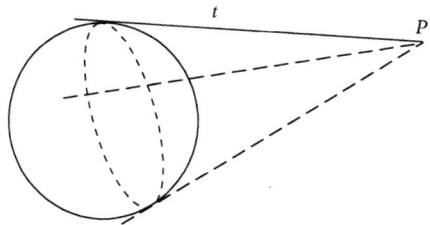

Figure 12.9

If we visualize the sphere and tangent t rotating about the axis through P and the centre of sphere, it becomes obvious that t will pass through every tangent through P, and that t's contact point will pass through every other tangent's contact point, so that the distance between contact point and P will be the same for all tangents through P.[4]

Discussion of these examples

The first example, a sequence of symbol manipulations proving left cancellation in group theory, clearly falls on the side of algebraic thinking. And the last example, visualizing rotation of a sphere and tangent about a fixed axis, clearly falls on the side of geometric thinking. But the middle two examples, grafting trees of associativity and amalgamating arrow diagrams, are not so easily classified.

One attitude to this situation is that there really is a definite dividing line, with each example falling on just one side, but we do not know where it lies. Presumably on this view we need to make some discovery about what makes a deliberate sequence of mental representations diagrammatic (geometric) rather than symbolic (algebraic), or the reverse. But what is the ground for confidence that there is a definite dividing line?

An alternative view, one that I find more plausible, is that there is a rough subjective distinction here, and our inability to classify the intermediate cases arises not from ignorance but indeterminacy. How can this be? One possibility is that there are several independent features whose presence or absence unconsciously affects our inclination to classify thinking as symbolic or diagrammatic. A clear symbolic case, such as the proof of left-cancellation, will have, say, features A, B, and C, but lack D, E, and F, while a clear diagrammatic case, such as the sphere and tangent, will have the opposite profile: it will *lack* A, B, and C, but *have* D, E, and F. If there are just 6 independent features there will be 64 possible profiles—the more independent features, the more profiles. Of those 64 profiles, 20 will be equidistant from the clearly symbolic and clearly diagrammatic profiles. With 8 independent features, 70 profiles will be equidistant from the clearly symbolic and clearly diagrammatic profiles. In general, relatively few profiles would seem to be clearly one or the other; the vast majority of profiles would cluster around the centre, having roughly as many features typical of symbolic cases as of diagrammatic cases.[5] So it would be no surprise at all that some examples of thinking are difficult to classify as symbolic or diagrammatic.

What sorts of features are involved? I do not know. This would be a task for cognitive science. And it may be a hard task: some of the features may not correspond to anything that we could easily make explicit; relevant features are unlikely to be equally weighted, and there may be some interdependence. As relevant features of the thinking are likely to depend on features of the representations themselves, it might be a useful start to list features that seem relevant to our classification of the representations. Some of these have already been mentioned. Here is the *kind* of thing I have in mind:

(1) Features typical of symbolic representations and untypical of diagrams:
- Discrete bits of information are carried by discrete parts of the representation;
- The representation has syntax;
- The representation carries only precise information; it does not carry vague information or mere suggestions.

(2) Features typical of diagrams and untypical of symbolic representations:
- Spatial relations between parts represent relations between what the parts represent;
- Information is presented simultaneously rather than serially;
- Some information is carried locally and some globally.

Since I came by these feature lists in an impressionistic way, without systematic empirical investigation, I have no confidence that these are the right features. Surely other features are relevant, and some or all of these may be irrelevant. They are purely illustrative. But taken in that spirit, the profiles of examples make my point.

> Example 1: the proof of left cancellation has all three features in the first list and none in the second.
>
> Example 6: the sphere and tangent of the ice cream cone proof has none of the features in the first list and all in the second.
>
> Example 4: The complex arrow diagram has all three features in the first list *and* all three features in the second list. So it straddles the diagrammatic-symbolic dichotomy.

Another preliminary task would be to consider the examples in the light of the known advantages and disadvantages of diagrammatic thinking as compared with symbolic thinking. A major advantage of diagrams is what Atsushi Shimojima calls *free rides*:[6]

Information that we would have to infer from a symbolic description of a situation can be simply "read off" a diagrammatic representation of that situation.

For example, from an amalgamation of commutative arrow diagrams we were able to read off an equality of arrows that was not among the original data. A major disadvantage of diagrams is what Shimojima calls *over-specificity*:[7]

> A situation cannot be represented as fulfilling a certain general condition by a diagram without also representing a *specific way* in which it is fulfilled, leading us to overlook possibilities that are unintentionally excluded.

In the example of the sphere and tangent, the point P is represented as lying at a distance from the sphere that is a more or less specific ratio to the sphere's diameter; it excludes ratios as large as 100 to 1 and as small as 1 to 2, but the exclusions are harmless in this case.[8] Notice, however, that while arrow diagrams can have the diagram's advantage of providing free rides, they do not suffer from the diagram's disadvantage of over-specificity. Once again, arrow diagrams defy the division into diagrammatic versus symbolic representations.

Arrow diagrams do not constitute an isolated case. Research journals contain papers with many kinds of visual representations that do not fall neatly into one of two exclusive classes, designated the symbolic (or algebraic) and the diagrammatic (or geometric). Here I reproduce four examples without discussion. Figure 12.10 is from a paper by Tom Leinster in *Theory and Applications of Categories*.[9] The second example, Figure 12.11, is from a paper by Magnus Jacobsson in *Algebraic and Geometric Topology*.[10] Figures 12.12 and 12.13 are taken from a paper by Aaron Lauda published on the web.[11]

The moral of the story so far

The foregoing considerations show that the symbolic-diagrammatic distinction (or the algebraic-geometric distinction) is too coarse to advance our understanding of visual thinking in mathematics, let alone mathematical thinking in general: we need taxa of finer grain.[12] Although there is a contrast to be drawn between symbolic thinking and diagrammatic

ALGEBRAIC V. GEOMETRIC THINKING 251

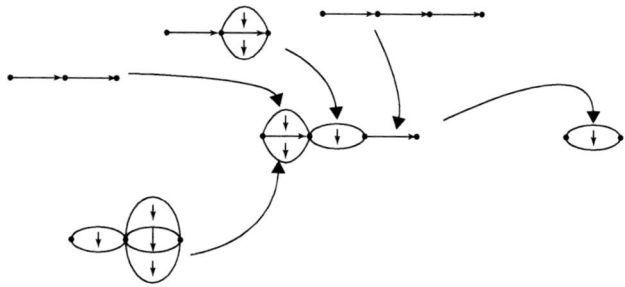

represents the composition

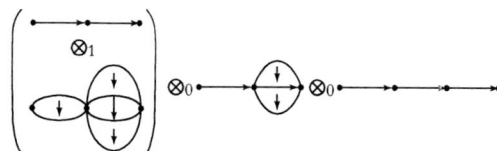

Figure 12.10

Figure 12.11

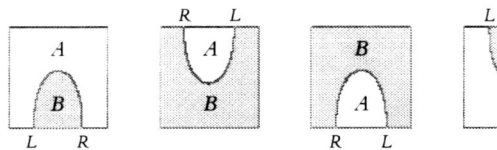

The zigzag laws for the four maps above are depicted in string notation as:

Figure 12.12

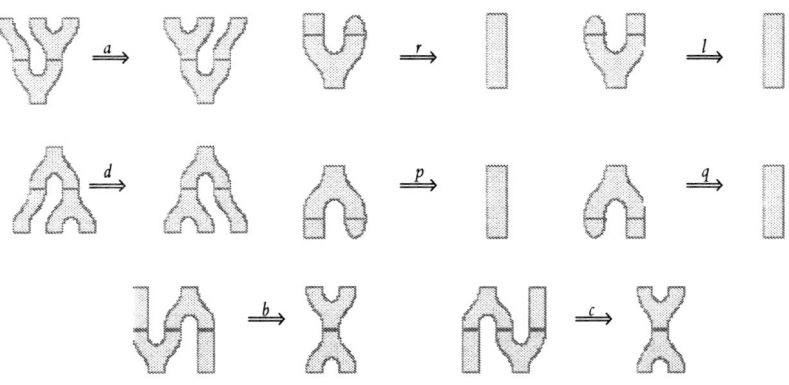

Figure 12.13

thinking, these are vague opposites that supervene on something like a lattice of possibilities. To think of the symbolic-diagrammatic contrast as constituting a division of all mathematical thinking into two kinds is no more apt than thinking that every human is wise or foolish.

Even if we restrict consideration to representations that we would not hesitate to regard as diagrams, there is diversity. Diagrams themselves belong to a spectrum, depending on the extent to which they depend on *resemblance* on the one hand and on *conventions* of representation on the other hand.[13] Arrow diagrams as used in category theory are highly conventional; they cannot rely on resemblance, because there are categories of many different kinds and often their elements are quite abstract. Graphs of functions rely largely on conventions, and a curve in the graph of a function almost never resembles what is described by the function. Geometrical diagrams also rely on conventions of course; but they rely to a significant extent on resemblance, which gives them their special utility in geometrical problem solving.

There may be many cognitively significant differences within the range of thinking considered here. For example, it is possible that much thinking that we would count as clearly symbolic uses a syntax processing capacity evolved for natural language, while other visual thinking does not. Is this true? If not, what are the cognitively significant links between thinking with symbolic formulas and natural language sentence processing? To get clear about the cognitively significant distinctions in mathematical thinking we must go beyond impressionistic classifications and seek to develop a taxonomy based on cognitive experiments.

The inadequacy of a binary distinction

Now I want to argue for a more aggressive claim: any division of mathematical thinking into just two kinds—algebraic-geometric, symbolic-diagrammatic, or whatever—is liable to be misleading. To see this, let us look at four ways of reaching the formula for the nth triangular number, that is, the sum of integers from 1 to n, which I will denote "$T(n)$":

$$T(n) = (n^2 + n)/2.$$

The Inductive argument The first is an inductive argument on the positive integers.

Basis. $T(1) = 1$, trivially. A half of $1^2 + 1$ is also 1. So the formula holds for $n = 1$.
Inductive hypothesis. $T(m) = (m^2 + m)/2$.
Then $T(m+1) = T(m) + m + 1$
$= (m^2 + m)/2 + m + 1$
$= (m^2 + m + 2m + 2)/2$
$= (m^2 + 2m + 1 + m + 1)/2$
$= ((m+1)^2 + (m+1))/2$.

Hence by induction the formula holds for all positive integers n.

The Gaussian argument The second argument is often found in textbooks. This argument is adapted from the quick and easy method Gauss used to find the sum of the integers from 1 to 100, when he was 10 years old.[14] Consider an array consisting of the numerals for the first n positive integers running from left to right in order, and, below this row, another row with the same numerals in reverse order, with vertical alignment of terms:

| 1 | 2 | 3 | .... | $n-2$ | $n-1$ | n |
| n | $n-1$ | $n-2$ | .... | 3 | 2 | 1 |

The first row sums to $T(n)$. So does the second row. Hence the terms in the whole array sum to $2 \times T(n)$. Each of the columns formed by these terms sums to $n + 1$ and there are n of these columns. Hence the terms in the whole array sum to $n \times (n + 1)$. Hence $2 \times T(n) = n \times (n + 1)$. Hence $T(n) = (n^2 + n)/2$.

The Pebble argument Here is an argument presented in Chapter 8, an ancient "pebble" argument. (But I shall talk of dots rather than pebbles.) The thinking applies only when n is greater than 1, the case for 1 being clear by substitution in the formula.

Think of each number k as represented by a column of k dots. Consider a sequence of n such columns, with a 1-dot column on the left, ascending by 1 with each step to the right, with a horizontal base line, as in Figure 12.14. Then $T(n)$ is the total number of dots in this array.
Now consider a copy of this array next to the original. Invert the copy and place it above the original, so that the n-dot column of the copy is vertically in line with the 1-dot column of the original, and the 1-dot

ALGEBRAIC V. GEOMETRIC THINKING   255

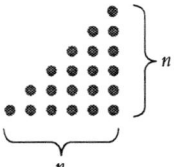

Figure 12.14

column of the copy is vertically in line with the n-dot column of the original. Then vertically close the two arrays until there is row alignment, as in Figure 12.15.

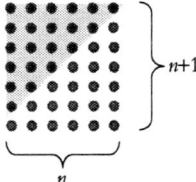

Figure 12.15

Clearly there are $2 \times T(n)$ dots in the whole array. Just as clearly, the array is a rectangle of n columns each consisting of $n + 1$ dots, making $n \times (n + 1)$ dots in all. Hence $2 \times T(n) = n \times (n + 1)$. Hence $T(n) = (n^2 + n)/2$.

The Square argument   Finally, there is an argument using areas. We use the argument for n greater than 1, the case for 1 being established by substitution in the formula. Consider a square of area n^2 units and divide it into unit squares. Now consider the lowest k unit squares in the kth column, going from the leftmost column to the right (shaded in Figure 12.16). Clearly $T(n)$ units is the total area of these unit squares taken together. Just as clearly, this area consists of half the total square area (under the diagonal from the lower left corner to the upper right) plus the total area of n half units: $n^2/2 + n/2$. Hence $T(n) = (n^2 + n)/2$.

Discussion of these four arguments

When we consider these four examples together, it seems clear where we would draw the line between symbolic and diagrammatic thinking: the Inductive argument and the Gaussian argument would count as symbolic, while the Pebble argument and the Square argument would count as

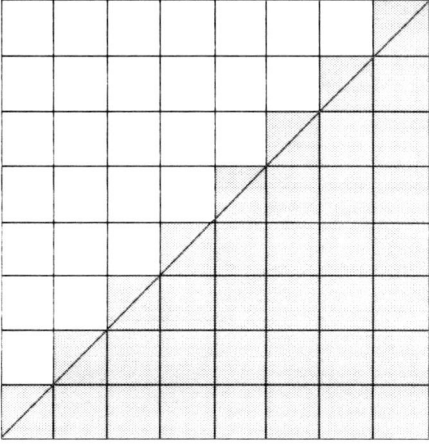

Figure 12.16

diagrammatic. Figure 12.17 may help us to keep all four cases in mind, and seems to reinforce the inclination to draw the symbolic-diagrammatic dividing line between the first pair and the last pair.

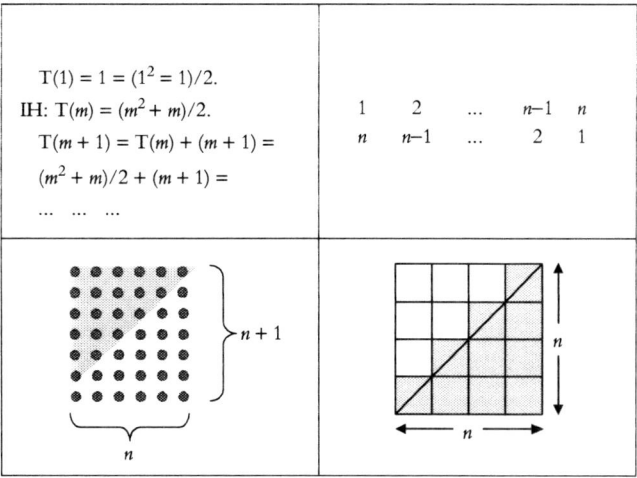

Figure 12.17

The Gaussian argument and the Pebble argument But this division obliterates significant structural parallels between the Gaussian argument and the Pebble argument:

(a) In following the Gaussian argument we must think of the array of numerals in two different ways, as a *pair of rows* and as *n columns*. In following the Pebble argument we must think of the array of dots in two different ways, as a *pair of triangular arrays* and as *n columns*.

(b) In both cases, the array consists of two parts, upper and lower, each consisting of a sequence of number representatives.

(c) In both cases, the numbers represented in one of these parts *decrease* by 1, each step from left to right, while the numbers represented in the other part *increase* by 1, each step from left to right, so that the sum of the numbers represented in each column is constant.

(d) The similarity in construction is matched by a similarity in reasoning. In both cases, we must recognize that there are n columns summing to $n + 1$, and infer that the total is $n \times (n + 1)$.

(e) In both cases, we must recognize that each part represents $T(n)$, and infer that the total is $2 \times T(n)$.

(f) In both cases, we infer that $2 \times T(n) = n \times (n + 1)$, and reach the formula for triangular numbers by dividing by 2.

The shared features of construction and reasoning in these two cases are so striking that one might regard them as two presentations of the same argument. At the same time, there is an important difference between them. To grasp the significance of the array in the Gaussian argument, one must know what the numerals mean, whereas no semantic knowledge is required to grasp the significance of the array in the Pebble argument. This is cognitively significant: a highly localized brain lesion could deprive you of your grasp of numerals thus preventing you from following the Gaussian argument, while your ability to follow the Pebble argument remained intact. So I am not advocating lumping the Gaussian thinking and the Pebble thinking together. My claim is just that it is quite misleading to regard them as falling either side of a fundamental divide.

This means that any acceptable way of dividing our four examples into just two classes must keep the Gaussian thinking and the Pebble thinking together. So if the four cases are divided into pairs, the thinking involved in following the Square argument must be classified together with the thinking involved in following the Inductive argument. I take it that this classification is too unattractive to merit discussion. That leaves only the possibility that one of the four cases is isolated from the rest, if we insist on a simple dichotomy, and the isolated case must be either the Square thinking or the Inductive thinking. That in turn means that one of these must be classified together with the Gaussian thinking and the Pebble thinking. But *that* will not run, I claim: the Square thinking and the Inductive thinking are each importantly different from all the others. To see this, let us look at each of them in turn.

The Square argument and the Inductive argument In following the Square argument a perceptual concept for a square will be activated. Something very similar is true of the thinking in following the Pebble argument—the relevant concept will be a perceptual concept for a rectangle. This is an important similarity, but it is quite superficial to classify the Pebble argument along with the Square argument on this basis. The thinking involved in following the Square argument is distinguished from the thinking involved in following each of the other three arguments in two ways. First, we must take the figure perceived as a square to represent a geometrical object, a geometrically perfect square composed of geometrically perfect unit squares. Secondly, we must assume a couple of geometrical facts:

(1) A square composed of n columns of n unit squares has area n^2 units;
(2) The area under a diagonal of a square is half the area of the square.

None of the other examples directs attention to a geometrical object; none of the other examples uses geometrical facts. So only in the case of the Square argument can the thinking involved be properly regarded as geometrical thinking. In this way the thinking of the Square argument is importantly different from the rest.

What about the Inductive argument? Thinking through the Inductive argument differs from the others in two major ways. One of these is that it does not require visually attending to an "external" spatial object, that is, a spatial object beyond the inscriptions of sentences or formulas used to express the argument. Although one has to attend to those sentences, that is because they express the relevant thoughts; it is not because the thoughts are *about* those sentences. In the other cases, following the argument does require visually attending to an external spatial object.

A second major way in which the Inductive thinking differs from the others lies in the way the generality of the conclusion is reached. Here generality is reached by using mathematical induction. In the other three arguments generality is reached by indicating a uniform method of constructing a spatial object with a numerical parameter $n > 1$, and by reasoning about the spatial object in a way that does not depend on supplying a value for n. Consider the construction of the rectangular array of dots in the Pebble argument. Although illustrated for $n = 6$, the same method of construction and the same reasoning can be used for all values of n greater than 1.[15] The reasoning has a schematic character. The tacit logic of the reasoning in these cases is:

Here is a sound argument *schema* with parameter n and conclusion $C(n)$. Hence for all n, $C(n)$.[16]

These two differences are epistemically as well as cognitively important. One is most unlikely to *discover* the formula for triangular numbers by constructing an inductive argument, as one must already have in mind a candidate formula for the nth triangular number in order to construct an argument for the inductive step.[17] What is going to bring a candidate formula to mind? Answer: schematic visual thinking of the kind illustrated by the other three cases. Ease of discovery is one of the advantages of visual thinking. A disadvantage is that we are liable to overgeneralize by assuming that the uniform method or schema works even for extreme or untypical cases. We need to check or prove the deliverances of visual thinking. That is where an inductive argument has the advantage: it provides the needed check.

So the Inductive thinking is importantly different from the visuo-spatial thinking of the other three cases. But if we draw the dividing line between the Inductive thinking and the rest, we would be ignoring obvious and important differences among the other three cases. The Square Argument is a genuinely geometrical argument. It relies on theorems of Euclidean plane geometry, and does not involve any arrangement of numerals or variables. The opposite is true of the Gaussian argument. Are not these the kind of differences that led to the geometric-algebraic classification in the first place? Even if that classification is specious, those differences are not insignificant, and there is no good reason to overlook them.

The clear lesson is that *no twofold division* of thinking in mathematics will suffice for a proper treatment of the thinking involved in following these four arguments. We need a taxonomy that is both more discriminating and more comprehensive. At least four kinds of thinking are exemplified by these cases. And if we reflect on the six examples discussed in the first section of this chapter and the examples discussed in earlier chapters, we may find many more. There is no problem with sharp twofold divisions. The point is that we will need a plurality of them, and the classes they separate will often intersect, resulting in many taxa, none of which coincide with any of the vague impressionistic categories we acquire unreflectively.

Operations in spatial thinking

What I hope will have come across from the book up to this point is the abundant diversity of visuo-spatial thinking in mathematics. This I think is the main arena of mathematical discovery and explanation, and the font of aesthetic pleasure in mathematics. Yet we still lack a useful taxonomy of spatial thinking in mathematics. To underline this, and to help see why spatial thinking is so useful for the purpose of discovery, I want now to consider kinds of spatial thinking some of which occur across the diagrammatic-symbolic spectrum. These are specific operations of spatial thought. I divide them into those that involve visualizing motion and those that do not. Among those that do not involve visualizing motion, two important operations are noticing reflection symmetries and aspect shifting. I will discuss these briefly.

Noticing reflection symmetry The ability to detect reflection symmetry, especially vertical reflection symmetry, appears early in development[18] and has a major role in shape perception.[19] Its importance in basic geometrical knowledge surfaced in Chapters 3 and 4, and there is a justly famous classical example: to make the discovery that the square on a diagonal of a given square has double the area in the manner suggested in Plato's *Meno*, we must see the small triangles as congruent, and this depends on our ability to detect certain reflection symmetries. This should be clear from Figure 12.18.

I do not want to exaggerate the importance of noticing reflection symmetries. But I suspect that it plays a large role in the cognitive foundations of geometry.

ALGEBRAIC V. GEOMETRIC THINKING   261

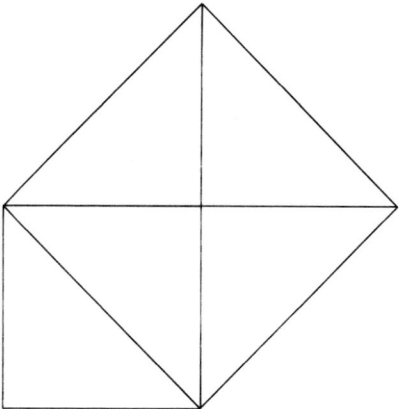

Figure 12.18

Aspect shifting   It is well known that presented with a rectangular array of black dots in columns and rows, if intercolumn spacing is not too different from inter-row spacing we can intentionally change from seeing the array as composed of columns to seeing it as composed of rows, and vice versa. In such cases one is exercising a capacity for visual aspect shift. There is some evidence that this can occur in visual imagination too. In one test, subjects were asked to do the following: visualize an upper case letter D, then to mentally rotate it 90° anticlockwise, then place an upper case letter J directly beneath it. Then they were asked what object the resulting image looked like. Subjects tended to identify it as an umbrella.[20] Visual aspect shift may be a kind of attentional shift; like other kinds of attentional change it can occur without conscious effort.

Several earlier examples involve aspect shifts. The thinking of the Gaussian argument involves viewing the array of symbols in two different ways: (1) as a pair of rows of addends and (2) as n columns of addends. Once the two-dimensional array is in place we do not rearrange anything. In the Pebble argument we are required to see the rectangular array of dots in two different ways: (1) as a pair of triangles of dots and (2) as n columns of dots. In thinking through the Square argument for the formula for triangular numbers, one views a specific portion of the square, the shaded portion, in two different ways: (1) as a sequence of columns of unit squares and (2) as the part of a square below a diagonal plus several half unit squares. No transformation of the figure is involved, no rearrangement of parts, no motion of the whole. There are plenty of other cases. The example from the *Meno* involves seeing one of the triangles as (1) a half of the small square and as (2) a quarter of the large square. Sometimes discovery in symbolic thinking involves seeing a single expression as an instance of two distinct forms, as suggested in Chapter 10. Aspect shifts are not uncommon in mathematical thinking and may be an important means of discovery.

Visualizing motions Visualizing motions has a large role in mathematical thinking. I do not mean motions in the geometric sense, but in the physical sense. Let me explain the difference. In geometry we identify motions that have the same result, regardless of path, velocity, and so on. Secondly, reflection symmetries, which are not physical motions, are regarded as motions in geometry. Finally, certain geometric mappings are taken to be motions of the whole space, isometries and dilatations. The idea of "motions of the whole space" can be puzzling. What we actually visualize in these cases, at least sometimes, is a uniform physical transformation of some objects *in* space, a transformation that we think of as applying universally, and we associate it with the mathematical description of a bijective mapping from the set of space-points to itself. One can think of isometries (mappings resulting from rigid motions) of the plane by imagining first a plane surface with objects marked on it and a copy on a transparency laid over it, and then moving the transparency in accord with a given rigid motion. In both cases what one is visualizing is physical movement of physical objects.

There are many kinds of visualized motion. Here are just some of them.

(1) A spatially bounded collection of objects moves rigidly, that is, with unchanging distances within and between objects. We use this kind of motion imagery, I suspect, in thinking of isometries.
(2) A spatially bounded collection of objects explodes uniformly from a central point O. This is how we imagine a *homothety*, also known as a central dilatation. This is a transformation under which, for some real number r each point P is mapped onto point P' such that O, P, P' are collinear and the distance between O and P' is r times the distance between O and P.

(3) A single object moves as a rigid whole, changing or not changing location. The rotating sphere and tangent mentioned earlier is an example.
(4) A single object undergoes a rearrangement of parts. A simple example is swapping terms flanking a binary operation symbol in applying a commutativity law. Another is the transformation of one associative tree into another by moving a component down one side of a "v" and up the other side.
(5) Continuous deformation, as in the famous remoulding of a coffee cup into a doughnut.
(6) Splits and fusions. Amalgamation of arrow diagrams is a kind of fusion.
(7) Others, e.g. the motions and intercalations visualized in matrix multiplication.

This is a preliminary, tentative, and incomplete classification. Empirical investigation might well find that other distinctions have greater cognitive importance.

These are just a few of the wide variety of visualized motions that occur in mathematical thinking. Cognitively speaking, they fall under the heading of image transformations. There is plenty of evidence about image transformations following the pioneering work of Shepard and his colleagues. This and later work is discussed at length by Kosslyn,[21] himself an outstanding contributor to that later work. There are cognitive distinctions among visualized motions that I have ignored. For example, sometimes a motion is visualized by recalling a seen motion of the relevant kind, such as birds flying in formation; at other times the motion is added rather than recalled, as when we imagine a coffee cup continuously deforming into a doughnut. That is Kosslyn's distinction between motion-encoded image transformations and motion-added image transformations. Visualized motions also divide into cases in which one imagines that the movement is brought about by oneself, and cases in which it is not so imagined. When I visualize a coffee cup deforming I do not imagine this as something I am causing. But when I visualize a cube sitting on a plane moved so that it stands on one vertex, I can imagine this as a motion that I am causing,

264 ALGEBRAIC V. GEOMETRIC THINKING

as if by a hand action. Even symbol movements can have a haptic feel, something perhaps reflected in the metaphor of symbol *manipulation*.

Conclusion

The epistemic value of these operations of visual imagination, and others like them, lies in the fact that they easily promote discovery by bringing to our attention new information. In the case of a reflection symmetry, equalities of angle or length or shape are made immediate to us, and information already acquired about the part of a figure on one side of an axis reproduces as information about the other part. Shifting aspect entails newly noticing some fact about an object already attended to. Visualizing a motion reveals that a certain operation on something has a certain result.

In general, visualizing provides an almost effortless way of acquiring new information, and its results often come with a degree of immediacy, clarity, and force that makes visualization apt as a means of discovery and explanation.[22] Thus the epistemic value of visual thinking in mathematics becomes obvious, once we give due importance to discovery and explanation. Of course there are pitfalls with visual thinking. That is why one needs a canon of principles for checking a claimed discovery, for proving it. The importance of proof has blocked epistemic interest in mathematical discovery for a century, when all eyes were on questions of logic, proof and justification—even to the extent that mathematics was widely held to be nothing but logic, and a major debate in philosophy of mathematics was boiled down to a dispute about the logical law of excluded middle. The balance needs to be redressed, with discovery, explanation, and application alongside proof and logic as topics of mathematical epistemology.[23]

Interest in visual thinking in mathematics is having a revival somewhat ahead of interest in mathematical discovery and explanation. In recent years there has been considerable effort in showing the efficacy of visual methods as means of proof.[24] In my judgement the results of these efforts are important for refuting a narrowly linguistic view of proof; but in other respects the results are meagre. Visual thinking is much more fruitful at the level of discovery. In earlier chapters I hope to have shown how genuine discovery by visual means is possible in mathematics, and to have lifted the lid enough to give some idea of the unrecorded wealth of visual thinking in mathematics. In this chapter I hope to have shown that the diversity of visual thinking in mathematics outruns any twofold classification. The indications of this preliminary study are that a valuable taxonomy of visual thinking in mathematics is likely to be complicated and may need several dimensions: it will take some time to emerge. Visual thinking in mathematics is extensive, diverse, familiar, and yet little understood. Here is abundant terrain for research. Contributions can be expected from mathematics education, history and philosophy of mathematics, and computer science; but above all I would look to, and advocate, research in the cognitive sciences. As yet we have barely started.

Notes

1. Frederic Patras introduced me to associahedra and their trees.
2. If you are not sure how to do that, just imagine the unit vector rotating anticlockwise, viewing it as the hypotenuse of the right-angled triangle whose other two sides are the projection of the vector onto the horizontal axis and the vertical segment connecting the vector's endpoint with the endpoint of its horizontal projection. As the vector rotates, the angle θ between the vector and its horizontal projection increases from 0 to π radians (= 180 degrees). The projection length is cos(θ), as the latter is "adjacent over hypotenuse"—the projection is the adjacent side while the hypotenuse has unit length. Now imagine what will happen to cos(θ), which is the projection length, as θ increases from 0 to π; and imagine what will happen to sin(θ), which is the length of the vertical segment joining endpoints (opposite over hypotenuse), as θ increases from 0 to π.
3. For substantiation, see the discussion about the possibility of visual discovery of Bolzano's theorem in Ch. 9.
4. This is part of Dandelin's "ice cream cone proof" that the definition of an ellipse in terms of conic sections entails that an ellipse is the set of points in a plane the sum of whose distances from two fixed points in that plane is constant. Apostol (1967: ch. 13).
5. For exactly 8 independent features, if we count as clearly symbolic or clearly diagrammatic also those profiles that differ by one feature from the paradigm profiles, and if we count as in-between also those that differ by one feature from a profile equidistant from the paradigms, just 9 profiles will be clearly symbolic, just 9 will be clearly diagrammatic, while 182 profiles will be in between.

6. Shimojima (2004a, 2004b). The article (2004a) is an abstract with bibliography for Shimojima's highly illuminating PowerPoint talk on diagrams (2004b).
7. Shimojima (2004a, 2004b). Norman (2006) has a nice way of putting this: compared with verbal descriptions diagrams typically *lack discretion*.
8. See the section headed "The problem of unintended exclusions" in Chapter 8 for discussion of the possible epistemic consequences of unintended exclusions.
9. Leinster (2004).
10. Jacobsson (2004).
11. Lauda (2006).
12. This way of putting it was suggested in discussion by B. Chandrasekeran.
13. David Tall pointed out to me that a spectrum is normally thought of as a linear ordering of its elements; so the metaphor of a spectrum occludes the possibility that the kinds of thinking under consideration are only partially ordered, if ordered at all.
14. Hall (1970) reports Gauss as thinking of the numerals from 1 to 50 in a row from left to right, then below them the numerals from 51 to 100 in a row from right to left, so that 51 occurs below 50, 52 below 49, and so on with 100 below 1 on the very left. The result is an array of 50 columns that sum to 101, giving 5050 in all.
15. Indeed, one can visualize the construction in such a way that there is no actual number k such that exactly k columns are represented in the image. Image indeterminacy is discussed at length in Chapter 8.
16. For corroboration of this view of the matter, see Jamnik (2001). "C(n)" here abbreviates "$n > 1 \to T(n) = (n^2 + n)/2$".
17. Inductive arguments often have a *base* case, which proves the condition for the initial element D(i), and an *inductive step*, which proves that if a number fulfils the condition so does its successor. In symbols: $\forall n(D(n) \to D(n+1))$. But some inductive arguments do without a base case by means of an inductive step that effectively incorporates it: if for all $k < n$ D(k), then D(n). In symbols: $\forall n(\forall k < n \, D(k) \to D(n))$. As no element precedes the initial element i, it is trivially true that $\forall k < i \, D(k)$, as this is equivalent to $\forall k \, [\neg(k < i) \text{ or } D(k)]$. So D($i$) follows. All inductive arguments have an inductive step.
18. Fisher et al. (1981).
19. See Palmer (1985) and Tyler's introduction to Tyler (1996).

20. Finke et al. (1989).
21. Shepard and Cooper (1982); Kosslyn (1994).
22. See Mancosu's valuable studies of explanation in mathematics: Mancosu (2000, 2001, and 2004).
23. See Giaquinto (2005b) for support and elaboration of this view.
24. Allwein and Barwise (1996); Brown (1999); Jamnik (2001).

Bibliography

Allwein, G., and Barwise, J. (eds.) (1996), *Logical Reasoning with Diagrams* (Oxford: Oxford University Press).
Apostol, T. (1957), *Mathematical Analysis* (Reading, Mass.: Addison-Wesley).
──── (1967), *Calculus*, 2nd edn. (New York: Wiley).
Atkinson, J., Campbell, F., and Francis, M. (1976), 'The magic number 4 ± 0: a new look at visual numerosity judgements', *Perception*, **5**: 327–34.
Attneave, F. (1962), 'Perception and related areas', in S. Koch (ed.), *Psychology: A Study of a Science*. (New York: McGraw-Hill).
Bachot, J., Gevers, W., Fias, W., and Roeyers, H. (2005), 'Number sense in children with visuospatial disabilities: orientation of the mental number line', *Psychology Science*, **47**: 172–83.
Bächtold, D., Baumüller, M., and Brugger, P. (1998), 'Stimulus-response compatibility in representational space', *Neuropsychologia*, **36**: 731–5.
Baird, J., and Noma, E. (1975), 'Psychophysical study of numbers. I: Generation of numerical response', *Psychological Research*, **37**: 281–97.
Banks, W., and Coleman, M. (1981), 'Two subjective scales of number', *Perception and Psychophysics*, **29**: 95–105.
Barwise, J., and Etchemendy, J. (1991a), 'Visual information and valid reasoning', in W. Zimmerman and S. Cunningham (eds.), *Visualizing in Teaching and Learning Mathematics* (Washington: Mathematical Association of America); repr. in Allwein and Barwise (1996).
──── ──── (1991b), 'Heterogeneous Logic', in J. Glasgow, N. Narayanan, and B. Chandrasekaran (eds.), *Diagrammatic Reasoning: Cognitive and Computational Perspectives* (Cambridge, Mass.: AAAI/MIT Press); repr. in Allwein and Barwise (1996).
──── and Hammer, E. (1994), 'Diagrams and the Concept of Logical System', in D. Gabbay (ed.), *What is a Logical System?* (Oxford: Clarendon Press); repr. in Allwein and Barwise (1996).
Baylis, G., and Driver, J. (1994), 'Parallel computation of symmetry but not repetition within single visual shapes', *Visual Cognition*, **1**: 377–400.
Berthoz, A. (2000), *The Brain's Sense of Movement*, trans. G. Weiss (Cambridge, Mass.: Harvard University Press); originally published as *Le Sens du mouvement* (Paris: Éditions Odile Jacob, 1997).
Biederman, I. (1995), 'Visual Object Recognition', in S. Kosslyn and D. Osherson (eds.), *Visual Cognition*, 2nd edn. (Cambridge, Mass.: MIT Press).

Bijeljac-Babic, R., Bertoncini, J., and Mehler, J. (1991), 'How do four-day-old infants categorize multisyllabic utterances?', *Developmental Psychology*, **29**: 711–21.

Bisiach, E., and Luzzatti, C. (1978), 'Unilateral neglect of representational space', *Cortex*, **14**: 129–33.

Block, N. (ed.) (1982), *Imagery* (Cambridge, Mass.: MIT Press).

—— (1983), 'The Photographic Fallacy in the Debate about Mental Imagery', *Noûs*, 17.

Bolzano, B. (1817), 'Purely analytic proof of the theorem that between any two values which give results of opposite sign there lies at least one real root of the equation', trans. S. Russ in W. Ewald (ed.), *From Kant to Hilbert: A Source Book in the Foundations of Mathematics* (Oxford: Oxford University Press, 1996).

Boysen, S. (1993), 'Counting in chimpanzees: Nonhuman principles and emergent properties of number', in S. Boysen and C. Capaldi (eds.), *The Development of Numerical Competence: Animal and Human Models* (Hillsdale, NJ: Lawrence Erlbaum).

Brannon, E., and Terrace, H. (1998), 'Ordering of the numerosities 1 to 9 by monkeys', *Science*, **282**: 746–9.

Brown, J. (1997), 'Proofs and Pictures', *British Journal for the Philosophy of Science*, **48**: 161–80.

—— (1999), *Philosophy of Mathematics: An Introduction to the World of Proofs and Pictures* (London: Routledge).

Bruce, V., and Morgan, M. (1975), 'Violations of symmetry and repetition in visual patterns', *Perception*, **4**: 239–49.

—— Green, P., and Georgeson, M. (1996), *Visual Perception: Physiology, Psychology, and Ecology*, 3rd edn. (Hove, Sussex: Psychology Press).

Bryant, P., (1974), *Perception and Understanding in Young Children: An Experimental Approach* (New York: Basic Books).

—— and Squire, S. (2001), 'Children's Mathematics: Lost and Found in Space', in M. Gattis (ed.), *Spatial Schemas and Abstract Thought* (London: MIT Press).

Burton, D. (1980), *Elementary Number Theory*, 2nd edn. (Boston: Allyn and Bacon).

Butterworth, B. (1999), *The Mathematical Brain* (London: Macmillan).

—— Zorzi, M., Girelli, L., and Jonckheere, A. (2001), 'Storage and retrieval of addition facts: The role of number comparison', *Quarterly Journal of Experimental Psychology*, **54A**: 1005–29.

Carnap, R. (1947), *Meaning and Necessity* (Chicago: University of Chicago Press).

Chambers, D., and Reisberg, D. (1985), 'Can mental images be ambiguous?', *Journal of Experimental Psychology: Human Perception and Performance*, **11**: 317–28.

Chi, M., and Klahr, D. (1975), 'Span and rate of apprehension in children and adults', *Journal of Experimental Child Psychology*, **19**: 434–9.

Cipolotti, L., and van Harskamp, N. (2001), 'Disturbances of number processing and calculation', in R. Berndt (ed.), *Handbook of Neuropsychology*, 2nd edn., vol. iii (Amsterdam: Elsevier Science).

Cohen, A., Ivry, R., Rafal, R., and Kohn, C. (1995), 'Activating response codes by stimuli in the neglected visual field', *Neuropsychology*, **9**: 165–73.

Cohen, L., and Ehrlich, G. (1963), *The Structure of the Real Number System* (Princeton: Van Nostrand).

Cooper, L., and Shepard, R. (1973), 'Chronometric studies of the rotation of mental image', in W. Chase (ed.), *Visual Information Processing* (New York: Academic Press); repr. in Shepard and Cooper (1982).

Corballis, M., and Roldan, C. (1975), 'Detection of symmetry as a function of angular orientation', *Journal of Experimental Psychology: Perception and Performance*, **1**: 221–30.

Crane, T. (1992), 'The non-conceptual content of experience', in T. Crane (ed.), *The Contents of Experience: Essays on Perception* (Cambridge: Cambridge University Press).

Dehaene, S. (1993), 'Varieties of numerical abilities', in S. Dehaene (ed.), *Numerical Cognition* (Oxford: Blackwell, 1993); originally published in 1992 in *Cognition*, **4**: 1–42.

——— (1997), *The Number Sense* (Oxford: Oxford University Press).

——— (2003), 'The neural basis of the Weber-Fechner law: a logarithmic mental number line', *Trends in Cognitive Sciences*, **7**: 145–7.

——— Dupoux, E., and Mehler, J. (1990), 'Is numerical comparison digital: Analogical and symbolic effects in two-digit number comparison', *Journal of Experimental Psychology: Human Perception and Performance*, **16**: 626–41.

——— and Mehler, J. (1992), 'Cross-linguistic regularities in the frequency of number words', *Cognition*, **12**: 1–29.

——— Bossini, S., and Giraux, P. (1993), 'The mental representation of parity and numerical magnitude', *Journal of Experimental Psychology: General*, **122**: 371–96.

——— and Cohen, L. (1994), 'Dissociable mechanisms of subitizing and counting—Neuropsychological evidence from simultanagnosic patients', *Journal of Experimental Psychology: Human Perception and Performance*, **20**: 958–75.

——— ——— (1995), 'Towards an anatomical and functional model of number processing', *Mathematical Cognition*, **1**: 83–120.

——— Naccache, L., Le Clerc'H, G., Koechlin, E., Mueller, M., Dehaene-Lambertz, G., van de Moortele, P., and Le Bihan, D. (1998), 'Imaging unconscious semantic priming', *Nature*, **395**: 597–600.

Deloche, G., and Seron, X. (eds.) (1987), *Mathematical Disabilities: A Cognitive Neuropsycholgical Perspective* (Hillsdale, NJ: Erlbaum).

DeValois, R. Morgan, H., and Snodderly, D. (1974), 'Psychophysical studies of monkey vision: III. Spatial luminance contrast sensitivity tests of macaque and human observers', *Vision Research*, **14**: 75–81.

Doricchi, F., Guariglia, P., Gasparini, M., and Tomaiuolo, F. (2005), 'Dissociation between physical and mental number bisection in right hemisphere brain damage', *Nature Neuroscience*, **8**: 1663–5.

Driver, J., Baylis, G., and Rafal, R. (1992), 'Preserved figure-ground segregation and symmetry detection in visual neglect', *Nature*, **360**: 73–5.

——and Mattingley, J. (1998), 'Parietal neglect and visual awareness', *Nature Neuroscience*, **1**: 17–22.

Duhem, P. (1914), *La Théorie physique: son objet, sa structure* (Paris: Riviere).

Ehrenwald, H. (1931), 'Störung der Zeitauffassung, der räumlichen Orientierung, des Zeichnens und des Rechnens bei einem Hirnverletsten', *Zeitschrift für die gesamte Neurologie und Psychiatrie*, **132**: 518–69.

Eisenberg, T. (1991), 'Functions and Associated Learning Difficulties', in D. Tall (ed.), *Advanced Mathematical Thinking* (Dordrecht: Kluwer Academic Publishers).

Euclid (1926), *The Thirteen Books of Euclid's Elements*, 2nd edn., 3 vols., trans. and commentary by T. Heath (Cambridge: Cambridge University Press); repr. by Dover Publications 1956.

Farah, M. (1985), 'Psychophysical evidence for a shared representational medium for visual images and percepts', *Journal of Experimental Psychology: General*, **114**: 91–103.

—— (1988), 'Is visual imagery really visual? Overlooked evidence from neuropsychology', *Psychological Review*, **95**: 305–17.

Fayol, M., Barrouillet, P., and Marinthe, C. (1998), 'Predicting arithmetical achievement from neuropsychological performance: A longitudinal study', *Cognition*, **68**: 63–70.

Ferrante, D., Gerbino, W., and Rock, I. (1995), 'Retinal vs. Environmental Orientation in the Perception of the Right Angle', *Acta Psychologica*, **88**: 25–32; repr. as *The Right Angle in Rock* (1997).

Finke, R. (1989), *Principles of Mental Imagery* (Cambridge, Mass.: MIT Press).

——, Pinker, S., and Farah, M. (1989), 'Reinterpreting visual patterns in mental imagery', *Cognitive Science*, **13**: 51–78.

Fischer, M., and Rottmann, J. (2005), 'Do negative numbers have a place on the mental number line?', *Psychology Science*, **47**: 22–32.

Fisher, C., Ferdinandsen, K., and Bornstein, M. (1981), 'The role of symmetry in infant form discrimination', *Child Development*, **51**: 457–62.

Foltz, G., Poltrock, S., and Potts, G. (1984), 'Mental comparison of size and magnitude: Size congruity effects', *Journal of Experimental Psychology: Learning, Memory and Cognition*, **10**: 442–53.

Fuson, K. (1988), *Children's Counting and Concepts of Number* (New York: Springer Verlag).

—— (1992), 'Relationships between Counting and Cardinality From Age 2 to Age 8', in J. Bideaud, C. Meljac, and J.-P. Fischer (eds.), *Pathways to Number* (Hillsdale, NJ: Lawrence Erlbaum Associates).

—— and Kwon, Y. (1992), 'Learning Addition and Subtraction: Effects of Number Words and Other Cultural Tools', in J. Bideaud, C. Meljac, and J.-P. Fischer (eds.), *Pathways to Number* (Hillsdale, NJ: Lawrence Erlbaum Associates).

Gallistel, C. (1990), *The Organization of Learning* (Cambridge, Mass.: MIT Press).

—— and Gelman, R. (1993), 'Preverbal and Verbal Counting and Computation', in S. Dehaene (ed.), *Numerical Cognition* (Oxford: Blackwell, 1993), originally published in 1992 in *Cognition*, **4**: 1–42.

Galloway, D. (1999), 'Seeing sequences', *Philosophy and Phenomenological Research*, **59**: 93–112.

Galton, F. (1880), 'Visualised numerals', *Nature*, **21**: 252–6; 494–5.

Garner, W. (1974), *The Processing of Information and Structure* (Potomac, Md.: Lawrence Erlbaum Associates).

Gelman, R., and Gallistel, C. (1978), *The Child's Understanding of Number* (Cambridge, Mass.: Harvard University Press).

Gentner, D. (1983), 'Structure-mapping: A theoretical framework for analogy', *Cognitive Science*, **7**: 155–70; repr. in A. Collins and E. Smith (eds.), *Readings in Cognitive Science: A Perspective from Psychology and Artificial Intelligence* (Palo Alto, Calif.: Kaufmann).

Gevers, W., Reynvoet, B., and Fias, W. (2003), 'The mental representation of ordinal sequences is spatially organised', *Cognition*, **87**: B87–B95.

Giaquinto, M. (1992), 'Visualizing as a Means of Geometrical Discovery', *Mind and Language*, **7**: 382–401.

—— (1993*a*), 'Diagrams: Socrates and Meno's slave', *International Journal of Philosophical Studies*, **1**: 81–97.

—— (1993*b*), 'Visualizing in Arithmetic', *Philosophy and Phenomenological Research*, **53**: 385–96.

—— (1994), 'Epistemology of Visual Thinking in Elementary Real Analysis', *British Journal for the Philosophy of Science*, **45**: 789–813.

—— (1996), 'Non-analytic conceptual knowledge', *Mind*, **105**: 249–68.

—— (1998), 'Epistemology of the Obvious: a geometrical case', in J. O'Leary Hawthorn (ed.), *A priori Knowledge, Philosophical Studies* (special issue), **92**: 181–204.

—— (2001), 'Knowing Numbers', *Journal of Philosophy*, **98**: 5–18.

—— (2002), *The Search for Certainty: A Philosophical Account of Foundations of Mathematics* (Oxford: Oxford University Press).

Giaquinto, M. (2005a), 'Symmetry Perception and Basic Geometrical Knowledge', in K. Joergensen and P. Mancosu (eds.), *Visualization, Explanation and Reasoning Styles in Mathematics* (Dordrecht: Kluwer Academic Publishers).

──── (2005b), 'Mathematical Activity', in K. Joergensen and P. Mancosu (eds.), *Visualization, Explanation and Reasoning Styles in Mathematics* (Dordrecht: Kluwer Academic Publishers).

Gödel, K. (1964), 'What is Cantor's Continuum Problem?', in Benacerraf and Putnam (eds.), *Philosophy of Mathematics: Selected Readings* (Cambridge: Cambridge University Press); revised and expanded from *American Mathematical Monthly*, **54** (1947), 515–25; errata **55**: 151; both versions are reprinted in Gödel (1990).

──── (1972), 'On an extension of finitary mathematics which has not yet been used', in Gödel (1990); a revised and expanded translation of 'Über eine bisher noch nicht benützte Erweiterung des finiten Standpunktes', *Dialectica*, **12** (1958), 280–7.

──── (1990), *Collected Works*, vol. ii (Oxford: Oxford University Press).

Goel, V., Buchel, C., Frith, C., Dolan, R. (2000), 'Dissociation of mechanisms underlying syllogistic reasoning', *NeuroImage*, **12**: 504–14.

Goldmeier, E. (1937), 'Similarity in visually perceived forms', *Psychological Issues*, **8**, Monograph 29 (1972); originally published in German.

Goodale, M. (1995), 'The cortical organization of visual perception and visuomotor control', in S. Kosslyn and D. Osherson (eds.), *Visual Cognition*, 2nd edn. (Cambridge, Mass.: MIT Press).

──── and Milner, A. (1992), 'Separate visual pathways for perception and action', *Trends in Neuroscience*, **15**: 20–5.

Hadamard, J. (1945), *The Psychology of Invention in the Mathematical Field* (Princeton: Princeton University Press); republished in 1954 by Dover Publications.

Hahn, H. (1933), 'The crisis in intuition', trans. in *Hans Hahn: Empiricism, Logic and Mathematics: Philosophical Papers*, ed. B. McGuiness (Dordrecht: D. Reidel 1980), first published in *Krise und Neuaufbau in den exakten Wissenschaften* (Leipzig and Vienna: Fünf Wiener Vorträge, 1933).

Hall, T. (1970), *Carl Friedrich Gauss: A Biography*, trans. A. Frodeburg (Cambridge, Mass.: MIT Press).

Hallett, D. (1991), 'Visualization and Calculus Reform', in W. Zimmermann and S. Cunningham (eds.), *Visualization in Teaching and Learning Mathematics* (Washington, DC: Mathematical Association of America).

Hallett, M. (1994), 'Hilbert's Axiomatic Method and Laws of Thought', in A. George (ed.), *Mathematics and Mind* (Oxford: Oxford University Press).

Halligan, P., and Marshall, J. (1988), 'How long is a piece of string? A study of line bisection in a case of visual neglect', *Cortex*, **24**: 321–8.

Harel, G., and Kaput, J. (1991), 'The Role of Conceptual Entities and Their Symbols in Building Advanced Mathematical Concepts', in D. Tall (ed.), *Advanced Mathematical Thinking* (Dordrecht: Kluwer Academic Publishers).

Hartje, W. (1987), 'The Effect of Spatial Disorders on Arithmetical Skills', in Deloche and Seron (eds.) (1987).

Hayes, J. (1973), 'On the function of visual imagery in elementary mathematics', in W. Chase (ed.), *Visual Information Processing* (New York: Academic Press).

Hershkowitz, R. (1987), 'The Acquisition of Concepts and Misconceptions in Basic Geometry', in D. Novak (ed.), *Proceedings of the 2nd International Seminar on Misconceptions and Educational Strategies in Science and Mathematics* (Ithaca: Cornell).

Hilbert, D. (1894), 'Die Grundlagen der Geometrie', ch. 2, in M. Hallett and U. Majer (eds.), *David Hilbert's Lectures on the Foundations of Geometry (1891–1902)* (Berlin: Springer, 2004).

—— (1925), 'Über das Unendliche', *Mathematische Annalen*, **95** (1926), 161–90; trans. in van Heijenoort (ed.), *From Frege to Gödel* (Cambridge, Mass.: Harvard University Press, 1967).

—— and Cohn-Vossen, S. (1952), *Geometry and the Imagination*, trans. P. Nemenyi (New York: Chelsea); originally published as *Anshaulich Geometrie* (Berlin: Springer, 1932).

Hinrichs, J., Yurko, D., and Hu, J. (1981), 'Two-digit number comparison: Use of place information', *Journal of Experimental Psychology: Human Perception and Performance*, **7**: 890–901.

Hinton, G. (1979), 'Some Demonstrations of the Effects of Structural Descriptions in Mental Imagery', *Cognitive Science*, **3**: 231–50.

—— and Parsons, L. (1981), 'Frames of reference and mental imagery', in J. Long and A. Baddeley (eds.), *Attention and Performance IX* (Hillsdale, NJ: Lawrence Erlbaum Associates).

—— —— (1988), 'Scene-based and viewer-centered representations for comparing shapes', *Cognition*, **30**: 1–35.

Horwich, P. (1998), *Meaning* (Oxford: Oxford University Press).

—— (2000), 'Stipulation', in P. Boghossian and C. Peacocke (ed.), *New Essays on the A Priori* (New York: Oxford University Press).

—— (2005), *Reflections on Meaning* (Oxford: Oxford University Press).

Hubel, D., and Wiesel, T. (1962), 'Receptive fields, binocular interaction and functional architecture in the cat's striate cortex', *Journal of Physiology*, **160**: 106–54.

—— —— (1968), 'Receptive fields and functional architecture of monkey striate cortex', *Journal of Physiology*, **195**: 215–43.

Hughes, A. (1959), *A History of Cytology* (London: Abelard-Schuman).

Humphreys, G. (1983), 'Reference frames and shape perception', *Cognitive Psychology*, **15**: 151–96.

Hurford, J. (2001), 'Languages Treat 1–4 Specially', *Mind and Language*, **16**: 69–75.

Hyman, I., and Neisser, U. (1991), *Reconstructing mental images: Problems of method* (Emory Cognition Project Technical Report 19, Emory University, Atlanta, Georgia).

Jacobsson, M. (2004), 'An invariant of link cobordisms from Khovanov homology', *Algebraic and Geometric Topology*, **4**: 1211–51.

Jamnik, M. (2001), *Mathematical Reasoning with Diagrams: From Intuition to Automation* (Stanford, Calif.: CSLI Publications).

Julesz, B. (1971), *Foundations of Cyclopean Perception* (Chicago: University of Chicago Press).

Kaan, E., and Swaab, T. (2002), 'The Brain Circuitry of Syntactic Processing', *Trends in Cognitive Sciences*, **6** (August), 350–5.

Kant, I. (1781–87), *The Critique of Pure Reason*, trans. and ed. P. Guyer and A. Wood (Cambridge: Cambridge University Press, 1998).

Kaplan, D. (1978), 'On the logic of demonstratives', *Journal of Philosophical Logic*, **8**: 89–98.

Kim, J. (1981), 'The role of perception in *a priori* knowledge', *Philosophical Studies*, **40**: 339–54.

Kinsbourne, M., and Warrington, E. (1962), 'A study of finger agnosia', *Brain*, **85**: 57–66.

——— (1963), 'The developmental Gerstmann syndrome', *Annals of Neurology*, **8**: 490–501.

Kitcher, P. (1984), *Mathematical Knowledge* (Oxford: Oxford University Press).

Klein, F. (1893), 'Vergleichende Betrachtungen über neurere geometrische Forschungen', *Mathematische Annalen*, **43**: 63–100 (Klein's 1872 Erlanger lecture).

Kline, M. (1972), *Mathematical Thought from Ancient to Modern Times* (New York: Oxford University Press).

Kosslyn, S. (1978), 'Measuring the visual angle of the mind's eye', *Cognitive Psychology*, **10**: 356–89.

——— (1980), *Image and Mind* (Cambridge, Mass.: Harvard University Press).

——— (1983), *Ghosts in the Mind's Machine* (New York: Norton).

——— (1994), *Image and Brain* (Cambridge, Mass.: MIT Press).

——— Koenig, O., Barrett, A., Cave, C., Tang, J., Gabrieli, J. (1989), 'Evidence for two types of spatial representations: Hemispheric specialization for categorical and coordinate relations', *Journal of Experimental Psychology: Human Perception and Performance*, **15**: 723–35.

Kosslyn, S., Thompson, W., and Alpert, N. (1997), 'Neural systems shared by visual imagery and visual perception: A positron emission tomography study', NeuroImage, **6**: 320–34.

Kundel, H., and Nodine, C. (1983), 'A visual concept shapes image perception', Radiology, **146**: 363–8.

Landau, E. (1934), *Differential and Integral Calculus*, trans. Hausner and Davis (New York: Chelsea, 1950).

Lauda, A. (2006), 'Frobenius Algebras and Planar Open String Topological Quantum Field Theories', http://arxiv.org/PS_cache/math/pdf/0508/0508349v1.pdf.

Leinster, T. (2004), 'Operads in Higher-Dimensional Category Theory', *Theory and Applications of Categories*, **12**: 73–194.

Lemmon, E. (1971), *Beginning Logic* (London: Nelson).

Lipton, J., and Spelke, E. (2003), 'Origins of number sense: Large number discrimination in human infants', *Psychological Science*, **14**: 396–401.

Littlewood, J. (1953), *Littlewood's Miscellany* (Cambridge: Cambridge University Press, 1986).

Lycan, W. (1986), 'Tacit Belief'. In R. Bogdan (ed.), *Belief* (Oxford: Clarendon Press).

McCloskey, M. (1992), 'Cognitive processes in numerical processing: evidence from acquired dyscalculia', *Cognition*, **44**: 107–57.

McDowell, J. (1994), *Mind and World* (Cambridge, Mass.: Harvard University Press).

—— (1998), Replies to Commentators, *Philosophy and Phenomenological Research*, **58**: 403–21.

Mach, E. (1897), *The Analysis of Sensations*, translated from the German edn. (New York: Dover, 1959).

MacLane, S. (1986), *Mathematics Form and Function* (Berlin: Springer-Verlag).

—— and Birkhoff, G. (1967) *Algebra* (London: Macmillan).

Mancosu, P. (2000), 'On Mathematical Explanation', in E. Grosholz and H. Breger (eds.), *Growth of Mathematical Knowledge* (Dordrecht: Kluwer Academic Publishers).

—— (2001), 'Mathematical Explanation: Problems and Prospects', *Topoi*, **20**: 97–117.

—— (2004), 'Visualization in mathematics', in K. Joergensen and P. Mancosu (eds.), *Visualization, Explanation and Reasoning Styles in Mathematics* (Dordrecht: Kluwer Academic Publishers).

—— and Hafner, J. (2004), 'The varieties of mathematical explanation', in K. Joergensen and P. Mancosu (eds.), *Visualization, Explanation and Reasoning Styles in Mathematics* (Dordrecht: Kluwer Academic Publishers).

Mandler, G., and Shebo, B. (1982), 'Subitizing: an analysis of its component processes', *Journal of Experimental Psychology: General*, **111**: 1–22.

Margolis, E., and Laurence, S. (1999), Introduction to E. Margolis and S. Laurence (eds.), *Concepts: Core Readings* (Cambridge, Mass.: MIT Press).

Marr, D., and Nishihara, H. (1978), 'Representation and Recognition of the spatial organization of three-dimensional shapes', *Proceedings of the Royal Society*, ser. B, **200**: 269–94.

Mechner, F. (1958), 'Probability relations within response sequences under ratio reinforcement', *Journal of the Experimental Analysis of Behaviour*, **1**: 109–22.

——— and Guevrekian, L. (1962), 'Effects of deprivation upon counting and timing in rats', *Journal of the Experimental Analysis of Behaviour*, **5**: 463–6.

Meck, W., and Church, R. (1983), 'A mode control model of counting and timing processes', *Journal of the Experimental Psychology: Animal Behaviour Processes*, **9**: 320–34.

Menninger, K. (1969), *Number Words and Number Symbols: A Cultural History of Numbers*, trans. P. Broneer (Cambridge, Mass.: MIT Press).

Metzler, J., and Shepard, R. (1974), 'Transformational Studies of the Internal Representation of Three-Dimensional Objects', in R. Solso (ed.), *Theories in Cognitive Psychology: The Loyola Symposium* (Hillsdale, NJ: Lawrence Erlbaum Associates), repr. in Shepard and Cooper (1982).

Mill, J. S. (1843), *A System of Logic Ratiocinative and Inductive* (London: John Parker).

Miller, N. (2001), 'A Diagrammatic Formal System for Euclidean Geometry', Ph.D. Thesis (Cornell University).

Moyer, R., and Landauer, T. (1967), 'Time required for judgements of numerical inequality', *Nature*, **215**: 1519–20.

——— ——— (1973), 'Determinants of reaction time for digit inequality judgements', *Bulletin of the Psychonomic Society*, **1**: 167–8.

Nakayama, K., He, Z., and Shimojo, S. (1995), 'Visual surface representation: A critical link between lower-level and higher-level vision', in S. Kosslyn and D. Osherson (eds.), *Visual Cognition*, 2nd edn. (Cambridge, Mass.: MIT Press).

Needham, T. (1997), *Visual Complex Analysis* (Oxford: Oxford University Press).

Neider, A., and Miller, E. (1993), 'Coding of Cognitive Magnitude: Compressed Scaling of Numerical Information in the Primate Prefrontal Cortex', *Neuron*: **37** 149–57.

Norman, J. (2006), *After Euclid* (Chicago: CSLI Publications, University of Chicago Press).

Nuerk, H., Weger, U., and Willmes, K. (2001), 'Decade breaks in the mental number line? Putting the tens and units back in different bins', *Cognition*, **82**: B25–B33.

Nuerk, H., and Willmes, K. (2005), 'On the magnitude representation of two-digit numbers', *Psychology Science*, **47**: 52–72.

O'Keefe, J. (1990), 'A computational theory of the cognitive map', *Progress in Brain Research*, **83**: 287–300.

―― (1991), 'The hippocampal cognitive map and navigational strategies', in J. Paillard (ed.), *Brain and Space* (Oxford: Oxford University Press).

―― and Nadel, L. (1978), *The Hippocampus as a Cognitive Map* (Oxford: Clarendon Press).

Page, J. (1993), 'Parsons on mathematical intuition', *Mind*, **102**: 223–32.

Palmer, S. (1980), 'What makes triangles point: Local and global effects in configurations of ambiguous triangles', *Cognitive Psychology*, **12**: 285–305.

―― (1983), 'The psychology of percepual organization: A transformational approach', in J. Beck, B. Hope, and A. Rosenfeld (eds.), *Human and Machine Vision* (New York: Academic Press).

―― (1985), 'The role of symmetry in shape perception', *Acta Psychologica*, **59**: 67–90.

―― (1990), 'Modern theories of gestalt perception', *Mind and Language*, **54**: 289–323.

―― (1999), *Vision Science: Photons to Phenomenology* (Cambridge, Mass.: MIT Press).

―― and Bucher, N. (1981), 'Configural effects in perceived pointing of ambiguous triangles', *Journal of Experimental Psychology: Human Perception and Performance*, **7** 88–114.

―― and Hemenway, K. (1978), 'Orientation and symmetry: Effects of multiple, rotational, and near symmetries', *Journal of Experimental Psychology: Human Perception and Performance*, **4**: 691–702.

Parsons, C. (1980), 'Mathematical intuition', *Proceedings of the Aristotelian Society*, NS **80** (1979–80), 145–68.

―― (1993), 'On some difficulties concerning intuition and intuitive knowledge', *Mind*, **102**: 232–46.

Peacocke, C. (1983), *Sense and Content* (Oxford: Clarendon Press).

―― (1992), *A Study of Concepts* (Cambridge, Mass.: MIT Press).

―― (1994), 'Nonconceptual Content: Kinds, Rationales and Relations', *Mind and Language*, **9**: 419–29.

―― (1998), 'Nonconceptual Content Defended', *Philosophy and Phenomenological Research*, **58**: 381–8.

Peacocke, C. (2001), 'Does Perception Have a Nonconceptual Content?', *Journal of Philosophy*, **98**: 239–64.

Pearce, J. (1997), *Animal Learning and Cognition: An Introduction*, 2nd edn. (Hove: Psychology Press).

Pesenti, M., Thioux, M., Seron, X., and De Volder, A. (2000), 'Neuroanatomical Substrates of Arabic Number Processing, Numerical Comparison, and Simple Addition: A PET Study', *Journal of Cognitive Neuroscience*, **12**: 461–79.

Piazza, M., Mechelli, A., Butterworth, B., and Price, C. (2002), 'Are Subitizing and Counting Implemented as Separate or Functionally Overlapping Processes?', *NeuroImage*, **15**: 435–46.

Plato (1985), *Plato: Meno*, ed. and trans. R. Sharples (Warminster: Aris and Phillips).

—— (1993), *Plato: Phaedo*, ed. C. Rowe (Cambridge: Cambridge University Press).

Podgorny, P., and Shepard, R. (1978), 'Functional representations common to visual perception and imagination', *Journal of Experimental Psychology: Human Perception and Performance*, **4**: 21–35.

Polya, G. (1962), *Mathematical Discovery* (New York: John Wiley and Sons).

Prior, A. (1960), 'The Runabout Inference-Ticket', *Analysis*, **21**: 38–9; repr. in P. Strawson (ed.), *Philosophical Logic* (Oxford: Oxford University Press, 1967).

Quine, W. (1951), 'Two Dogmas of Empiricism', repr. in W. Quine, *From a Logical Point of View* (Cambridge, Mass.: Harvard University Press, 1980); originally published in *Philosophical Review* (Jan. 1951).

—— (1960), 'Carnap and Logical Truth', repr. *Synthese*, **12** (1960); in W. Quine, *The Ways of Paradox and Other Essays* (Cambridge, Mass.: Harvard University Press, 1976).

Quinlan, P. (1996), 'Evidence for the use of scene-based frames of reference in two-dimensional shape recognition', in C. Tyler (ed.), *Human Symmetry Perception and its Computational Analysis* (Utrecht: VSP).

Resnik, M. (1997), *Mathematics as a Science of Patterns* (Oxford: Clarendon Press).

Robertson, I., and Marshall, J. (1993), *Unilateral Neglect: Clinical and experimental studies* (Hove: Lawrence Erlbaum Associates).

Rock, I. (1973), *Orientation and Form* (New York: Academic Press).

—— (1990), 'The concept of reference frame in psychology', in I. Rock (ed.), *The Legacy of Solomon Asch: Essays in Cognition and Social Psychology* (Hillsdale, NJ: Erlbaum).

—— (1997), *Indirect Perception* (Cambridge, Mass.: MIT Press).

—— and Leaman, R. (1963), 'An Experimental Analysis of Visual Symmetry', *Acta Psychologica*, **21**: 171–83; repr. as *Symmetry in Rock* (1997).

Rourke, B. (1993), 'Arithmetic disabilities, specific and otherwise: A neuropsychological perspective', *Journal of Learning Disabilities*, **26**: 214–26.

Rumbaugh, D., and Washburn, D. (1993), 'Counting by chimpanzees and ordinality judgments by macaques in video-formatted tasks', in S. Boysen and E. Capaldi (eds.), *The Development of Numerical Competence: Animal and Human Models* (Hillsdale, NJ: Lawrence Erlbaum).

Russell, B. (1912), *The Problems of Philosophy* (Oxford: Oxford University Press); repr. (Indianapolis: Hackett Publishing Company, 1990).
Sagan, H. (1994), *Space-Filling Curves* (New York: Springer-Verlag).
Salmon, N. (1988), 'How to Measure the Standard Metre', *Proceedings of the Aristotelian Society*, NS **88** (1987–8), 193–217.
Schwarz, W., and Keus, I. (2004), 'Moving the eyes along the mental number line: Comparing SNARC effects with saccadic and manual responses', *Perception & Psychophysics*, **66**: 651–64.
Schweinberger, S., and Stief, V. (2001), 'Implicit perception in patients with visual neglect: Lexical specificity in repetition priming', *Neuropsychologia*, **39**: 420–9.
Scott, S., Barnard, P., and May, J. (2001), 'Specifying executive representations and processes in number generation tasks', *Quarterly Journal of Experimental Psychology: Human Experimental Psychology*, **54A**: 641–64.
Segal, S., and Fusella, V. (1970), 'Influence of imaged pictures and sounds on detection of visual and auditory signals', *Journal of Experimental Psychology*, **83**: 458–64.
Seron, X., Pesenti, M., and Noël, M. (1992), 'Images of numbers, or "When 98 is upper left and 6 sky blue"', *Cognition*, **44**: 159–96; repr. in Dehaene (1993).
Shepard, R., and Feng, C. (1972), 'A Chronometric Study of Mental Paper Folding', *Cognitive Psychology*, **3**: 228–43; repr. in Shepard and Cooper (1982).
―― and Cooper, L. (eds.) (1982), *Mental Images and Their Transformations* (Cambridge, Mass.: MIT Press).
―― and Metzler, D. (1971), 'Mental Rotation of Three-dimensional Objects', *Science*, **171**: 701–3.
Shiffrin, R., McKay, D., and Shaffer, W. (1976), 'Attending to Forty-Nine Spatial Positions at Once', *Journal of Experimental Psychology: Human Perception and Performance*, **21**: 14–22.
Shimojima, A. (2004a), 'Inferential and Expressive Capacities of Graphical Representations: Survey and Some Generalizations', in A. Blackwell,, K. Marriott, and A. Shimojima (eds.), *Diagrammatic Representation and Inference* (Berlin: Springer).
―― (2004b), http://www.jaist.ac.jp/~ashimoji/Diagrams_2004_tutorial.ppt
Shin, S.-J. (1996), 'Situation-Theoretic Account of Valid Reasoning with Venn Diagrams', in Allwein and Barwise (1996).
Siegler, R., and Opfer, J. (2003), 'The development of numerical estimation: evidence for multiple representations of numerical quantity', *Psychological Science*, **14**: 237–43.
Simon, T. (1997), 'Reconceptualizing the origins of number knowledge: a non-numerical account', *Cognitive Development*, **12**: 349–72.

Simon, T. and Vaishnavi, S. (1996), 'Subitizing and counting depend on different attentional mechanisms: Evidence from visual enumeration in afterimages', *Perception and Psychophysics*, **58**/6: 915–26.

Starkey, P., Spelke, E., and Gelman, R. (1990), 'Numerical abstraction by human infants', *Cognition*, **36**: 97–128.

Sunderland, A. (1990), 'The bisected image? Visual memory in patients with visual neglect', in D. M. P. Hampson and J. Richardson (eds.), *Imagery: Current Developments* (London: Routledge).

Sutton, W. (1902), 'On the Morphology of the Chromosome Group in Brachystola Magna', *Biological Bulletin*, **4**: 24–39.

Tall, D. (1991), 'Intuition and Rigour: The role of Visualization in the Calculus', in W. Zimmeramnn and S. Cunningham (eds.), *Visualization in Teaching and Learning Mathematics* (Washington, DC: Mathematical Association of America).

Tappenden, J. (2001), 'Recent Work in Philosophy of Mathematics', *Journal of Philosophy*, **98**/9: 488–97.

Thompson, W., and Kosslyn, S. (2000), 'Neural systems activated during visual mental imagery: A review and meta-analyses', in A. Toga and J. Mazziotta (eds.), *Brain Mapping: The Systems* (London: Academic Press).

Thorpe, W. (1963), *Learning and Instinct in Animals* (London: Methuen).

Treisman, A. (1988), 'Features and Objects: The Fourteenth Bartlett Memorial Lecture', *Quarterly Journal of Experimental Psychology*, **40**A: 201–37.

—— Vieira, A., and Hayes, A. (1992), 'Automaticity and pre-attentive processing', *American Journal of Psychology*, **105**: 341–62.

Trick, L., and Pylyshyn, Z. (1991), 'A theory of enumeration that grows out of a general theory of vision: Subitizing, counting and FINSTs', University of Western Ontario COGMEM # 57.

—— —— (1994), 'Why are small and large numbers enumerated differently? A limited capacity preattentive stage in vision', *Psychological Review*, **100**: 80–102.

Tye, M. (1991), *The Imagery Debate* (Cambridge, Mass.: MIT Press).

—— (1993), 'Image indeterminacy: The picture theory of images and the bifurcation of "what" and "where" information in higher-level vision', in N. Eilan, R. McCarthy, and B. Brewer (eds.), *Spatial Representation* (Oxford: Blackwell).

Tyler, C. (ed.) (1996), *Human Symmetry Perception and its Computational Analysis* (Utrecht: VSP).

Uller, C., Carey, S., Huntley-Fenner, G., and Klatt, L. (1999), 'What representations might underlie infant numerical knowledge?', *Cognitive Development*, **14**: 1–36.

Ullman, S. (1996), *High-level Vision* (Cambridge, Mass.: MIT Press).

Ungerleider, L., and Mishkin, M. (1982), 'Two cortical visual systems', in D. Ingle, M. Goodale, and R. Mansfield (eds.), *Analysis of Visual Behaviour* (Cambridge, Mass.: MIT Press).
van Atten, M. (2002), *On Brouwer* (London: Wadsworth).
van Loosbroek, E., and Smitsman, A. (1990), 'Visual perception of numerosity in infancy', *Developmental Psychology*, **26**: 916–22.
van Oeffelen, M., and Vos, P. (1982), 'A probabilistic model for the discrimination of visual number', *Perception and Psychophysics*, **32**: 163–70.
Vinner, S. (1982), 'Conflicts between definitions and intuitions—The case of the tangent', in A. Vermandel (ed.), *Proceedings of the 6th International Conference for the Psychology of Mathematics Education* (Antwerp, Belgium: Universitaire Instelling Antwerpen).
—— and Hershkowitz, R. (1983), 'On Concept Formation in Geometry', *Zentralblatt fur Didaktik der Mathematik*, **15**: 20–5.
Vos, P., van Oeffelen, M., Tibosch, H., and Allik, J. (1988), 'Interactions between area and numerosity', *Psychological Research*, **50**: 148–54.
Wagemans, J. (1996), 'Detection of visual symmetries', in Tyler (1996).
Warrington, E. (1982), 'The fractionation of arithmetical skills: a single case study', *Quarterly Journal of Experimental Psychology*, **34**: 31–51.
Welford, A. (1960), 'The measurement of sensory-motor performance: Survey and reappraisal of twelve years' progress', *Ergonomics*, **3**: 189–230.
Wynn, K. (1995), 'Origins of numerical knowledge', *Mathematical Cognition*, **1**: 36–60.
Xu, F. (2003), 'Numerosity discrimination in infants: evidence for two systems of representation', *Cognition*, **89**: B15–B25.
—— and Spelke, E. (2000), 'Large number discrimination in 6-month-old infants', *Cognition*, **74**: B1–B11.
Zabrodsky, H., and Algom, D. (1996), 'Continuous symmetry: a model for human figural perception', in Tyler (1996).
Zago, L., Pesenti, M., Mellet, M., Crivello, F., Mazoyer, B., and Tzourio-Mazoyer, N. (2001), 'Neural correlates of simple and complex mental calculation', *NeuroImage*, **13**: 314–27.
Zebian, S. (2005), 'Linkages between Number Concepts, Spatial thinking, and Directionality of Writing: The SNARC Effect and the REVERSE SNARC Effect in English and Arabic Monoliterates, Biliterates, and Illiterate Arabic Speakers', *Journal of Cognition and Culture*, **5**: 165–91.
Zeki, S. (1993), *A Vision of the Brain* (Oxford: Blackwell).

Zorzi, M., and Butterworth, B. (1999), 'A computation model of number comparison', in M. Hahn and S. Stoness (eds.), *Proceedings of the Twenty First Annual Conference of the Cognitive Science Society* (Mahwah, NJ: Erlbaum), 778–83.

—— Priftis, X., and Umiltà, C. (2002), 'Neglect disrupts the mental number line', *Nature*, **417**: 138.

Index

algebra
  boolean 207–8, 223–4
  matrix 208, 210–11
  power-set 223–6
  vector 194–5
aspect shift 200, 259–61
Attneave, F. 106
axiom(s) 5–7, 12, 81, 194–202, 228
  Archimedean 189
  complete ordered field 231
  Dedekind-Peano 215
  Euclidean, *see also* parallels postulate 40

belief-forming dispositions 12, 35, 37, 39–42, 44–7, 61–4, 178–9
Bolzano 3, 169–71, 173, 177, 180–1, 190
Butterworth, B. 93, 95.

Carnap, R. 6
category pattern, *see* category specification
category specification 17, 23–4, 26–9, 37, 39, 43, 46, 60–2, 108–10, 112, 151–2, 154, 220–1, 227–8, 230, 232, 235–6
concepts 24–6, 35, 39, 45–6, 63, 111, 123, 154, 170, 174, 191
  analytic 170, 178–80, 182–3
  arithmetical 138, 156–7, 191,
  geometrical 12–13, 24, 26, 28–9, 35–8, 41–3, 46, 56, 61, 178–80, 229
  perceptual 24, 26–9, 46, 154, 178–80, 184, 229, 257
continuous function 3, 164, 170–3, 177–82
  nowhere-differentiable 3
curve 3, 99, 106, 164–74, 176, 179–83, 187–90, 252
  space-filling 4
Cramer's rule 192

Dedekind, R.
  Dedekind-continuity 229, 231
  Dedekind-Peano axioms 6, 215

Dehaene, S. 97, 98, 105–106,
diagrams 1, 8, 14, 16, 67, 113, 208, 217, 223–5, 242, 247, 249–50, 252
  in analysis 163, 177, 180, 182,
  arrow 243–6, 249–50, 252, 262
  in geometric proof 71–86
  thinking with, *see* thinking, diagrammatic
discovery 2, 5, 12, 18, 50, 64, 67, 71, 157–8, 211, 258–9, 261, 263–4
  geometrical 12, 50
  in analysis 163–4, 169, 172–4, 177, 180, 183–4, 190
  of general theorems 146, 148–9
Doricchi, F. 103, 107
Duhem, P. 44

Euclid's *Elements* 72
Euclidean geometry 74, 84, 84, 163, 178, 259

finitism 5
formal system 84, 194–6
free rides 249–50

Gallistel, C. 105
Gauss, C. 191, 253, 255–7
Gelman, R. 105
generalization 66, 77–80, 83–6, 88–9, 205, 208, 211
  inductive 43, 53, 63, 78, 125
  unwarranted 71, 77–8, 173, 182, 258
  from a diagram 80–1, 83
Gödel, K. 6, 231–3
graph 3, 168, 172, 180, 183, 264–7, 252

Hals, F. 144
Hayes, J. 133
Hilbert, D. 4, 7–8, 231–2
holistic empiricism 6
Horwich, P. 45

286 INDEX

imagery, *see* images
images 1, 7, 13, 59–62, 66, 99–103, 107–110, 112–15, 128, 131–4, 138, 142–4, 146, 150–4, 156–8, 166–7, 173, 182, 183, 203, 260–2
intuition 3, 6, 10, 123, 173, 183
intuitionism 5

Jacobsson, M. 250

Kanizsa figure 13
Kant 5–9, 40, 47, 68, 123–4, 126
knowledge 1, 5, 8–9, 17, 40, 42–3, 46, 47, 62, 67–8, 71, 84, 113, 124–7, 135, 190, 207, 210, 215–16
 a priori 12, 44, 47, 68, 127
 basic 35
 conceptual 45, 158
 empirical 123–4, 126
 finger 123
 geometrical 12, 29, 35, 40, 46, 50, 67–8, 259
 arithmetical 122–3, 125
 of structure 216–17, 224–6, 228, 231–3
 synthetic *a priori, see a priori*
Kosslyn, S. 17, 60, 108–9, 151–2, 154, 203, 262

Landau, E. 163, 177
Lauda, A. 250
left-cancellation law 242
Leinster, T. 250
limit
 cognitive 28, 83, 91, 153,
 mathematical 3–4, 7, 81, 172–4, 176, 187
Littlewood, J. 163, 177, 180, 184, 186–7

Mach, E. 15, 19, 20, 22
matrix 192, 208–211, 242, 245, 262
Mill 5, 9, 125
Miller, E. 106
Miller, N. 84
model, of a theory 215–6, 228, 231, 236

Nieder, A. 106
number line 8, 90–116, 128, 131, 134, 226–8

over-specificity, *see* unintended exclusions

Palmer, S. 19–20
parallels postulate 44–5
particularity objection 141–2, 145
Peacocke, C. 25
Peano, G. 4, 6, 215
pebble argument 159, 254–8, 261
Plato 5, 7, 12, 60, 112, 260
predicativism 5
Priestley, J. 45
proof 1–2, 5, 8, 10, 51, 67, 71–4, 77–8, 80–1, 83–6, 113–5, 123, 154, 163, 165–6, 170, 177, 180, 182–4, 194, 196, 211, 263
 of left-cancellation 242, 248–9
 ice-cream cone 249, 264
 of the infinity of primes 72, 87–8

Quine 6, 44

reflection symmetry 247, 263
Resnik, M. 217
Rock, I. 15, 16

schema 76, 154, 258
Scott, S. 106
shape perception 13, 15, 17
 and frame effects 18–19
 and intrinsic axes 18–19
 and reflection symmetry 19, 21, 28, 62, 259
Shepard, R. 18, 151
Shimojima, A. 249–50
space
 Euclidean 40, 52
structure 9, 204–8, 210–11, 214–36, 242–3
substitution 72, 154, 195, 197, 200–4, 254
symbol manipulation 7, 84, 134, 191–211
 concept-driven *v.* rule-driven 191–3
 formal *v.* informal 193–4

Theorem
 Bolzano's 169–73
 Pythagoras' 112, 240

Rolle's 164–9,
thinking
 algebraic, *see* symbolic
 diagrammatic 71, 73–4, 77–8, 84,
  241–2, 248–50, 252, 255
 geometric, *see* symbolic
 symbolic 9, 116, 132, 182–3, 191, 205,
  207–8, 211, 240–2, 248–50, 252–3,
  255, 261
trees
 binary 225–6,
 of associativity 243, 248

unintended exclusions 141, 145–6, 148,
 149, 150, 159, 250

visualizing
 geometrical discovery by 50–67
 in analysis 9, 163–84
 in arithmetic 121–34, 137–59
 motion 203–4, 209, 242, 259, 261–3
 number line 100, 107, 134
 numerals 99–101, 108
 rotation 18, 112, 151, 156, 201, 211,
  242–3, 248, 260
 scanning 109–10, 143–4, 225–7
 spatial transformation 219, 224
109–10, 143–4, 172, 225
 161, 217–224

Zorzi, M. 95, 101, 107